Meetings in English

Lisa Förster, Annette Joyce

Contents

Vorwort

Wer Geschäftskontakte mit internationalen Partnern pflegt, nimmt früher oder später an Meetings auf Englisch teil. Sind Sie auch in dieser Situation oder wollen Sie sich schon einmal vorsorglich fit machen? Dann sind Sie mit diesem TaschenGuide auf der sicheren Seite. Denn Meetings und Verhandlungen können in Ihrer Muttersprache schon schwierig genug sein, umso mehr in der Fremdsprache.

Wir zeigen Ihnen, wie Sie in Meetings auf Englisch durch Kompetenz, Verlässlichkeit und Höflichkeit Vertrauen aufbauen und überzeugen – egal, ob Sie das Meeting selbst leiten oder nur daran teilnehmen. Für alle Situationen eines Meetings – von der Begrüßung über die eigentlichen Verhandlungen bis zum Protokoll – geben wir Ihnen die notwendigen sprachlichen Mittel an die Hand und zeigen Ihnen kulturelle Besonderheiten.

Wenn Sie weniger aktive Erfahrung mit der englischen Sprache haben, ermöglicht Ihnen das Buch einen selbstbewussten Auftritt. Aber selbst Fortgeschrittene finden neue Ausdrücke und Wendungen, die ihren Sprachschatz sinnvoll ergänzen. Sie können es komplett durcharbeiten oder zwischendurch als Nachschlagewerk benutzen. Und um einen zusätzlichen Trainingseffekt zu erzielen, ist der gesamte TaschenGuide auf Englisch geschrieben. Wir wünschen Ihnen in Ihren künftigen Meetings viel Erfolg!

Lisa Förster und Annette Joyce

Preparing a meeting

In the run-up to a meeting there are a lot of things to bear in mind, especially when you are the one organising it. Booking a meeting room is only one task among many.

In this chapter, you will see how to

- invite people (page 6) and make arrangements (page 8),
- accept, refuse and postpone meetings (page 11),
- make the agenda (page 15),
- organise the meeting (page 21).

Inviting people to a meeting

In formal business relationships, you may wish to write a letter or email to a business associate to suggest an initial meeting. This correspondence can then be followed up with a telephone call or an email to confirm the meeting time and place.

Suggesting a meeting

In an established business relationship, a less formal style can be adopted for suggesting meetings by both telephone and email. Email is especially useful if a number of participants are involved. In a formal email, the style of language, salutation and complimentary close is the same as that used in a letter.

Example: formal UK-style letter/email

Marketing proposal: initial meeting

Dear Mr Smith

Thank you for your telephone call of this morning. I am pleased to enclose [letter]/attach [email] the information you requested and would welcome the opportunity to meet you in person to discuss the proposal and to answer any questions you may have.

I will contact you by telephone in the next few days to arrange a time that is convenient for you.

In the meantime, please do not hesitate to contact me if you require any further information.

Yours sincerely

David Braun

Business Development Manager

> **The correct way of writing salutations and endings**
> In **British English** letters and emails, the trend is to omit punctuation in the salutation, people's names (Mr, Mrs, Ms, Dr, etc) and in the complimentary close. Letters and emails end with "Yours sincerely", when the person is addressed by name, or "Yours faithfully", when the letter begins "Dear Sir or Madam". In **US English**, a period (British: full stop) follows abbreviations and the salutation ends with a colon, as in "Dear Ms. Jones:". Standard complimentary closes are "Sincerely yours," and "Sincerely," (note that both end with a comma).

Useful phrases

Less formal style

- I was wondering if we could meet in the near future to discuss ...?
- Perhaps it would speed things up if we met face to face to discuss this?
- Shall we meet next week to discuss the details in person?

Responding to a request for a meeting

Useful phrases

- I'm afraid my schedule is very full in the next few weeks. Would it be possible for us to discuss the matter by telephone instead?
- As we're both very busy, I would like to try to resolve the matter by telephone, if possible.
- I think it would be a good idea for us to meet.
- I agree that it would be beneficial for us to meet face to face.

Useful vocabulary

initial: erste(r, s)

complimentary close: Schlussformel

proposal: Vorschlag

to welcome an opportunity: eine Gelegenheit gern wahrnehmen

convenient: angenehm, bequem

schedule: Terminkalender

to resolve a matter: eine Sache klären

Making meeting arrangements

In an informal setting, once you have agreed with your business partner that you would like to meet, arrangements can be made by telephone or email. In both cases, the language used for organising the meeting is informal yet polite.

Who would like to meet when?

If you are organising a meeting with a larger number of participants and have to find out about their general availability on certain dates, "Doodle" can help. This is a clever tool you'll find online free of charge at www.doodle.ch. You create your doodle by just entering the dates and/or times you would like to suggest. A link to this doodle is automatically sent to your email account. You can then forward the link to all the prospective participants, asking them to state when they would be available and when not. There is also a field for comments.

Lunch meetings

If you meet over lunch make sure you can still take notes on a small pad. Follow your host's lead with regard to drinking alcohol. In Great Britain drinking alcohol in moderation is generally acceptable over lunch or dinner with business colleagues. However, if your business partners are sticking to water and you want to keep a clear head, avoid alcohol altogether.

Useful phrases

When?
- Would 20 November suit you?
- What does your schedule look like on 3 December?
- I'm afraid I'm away on business for the whole of that week. How about 25 November?
- What would be a convenient time for you?
- What time would suit you best?

Where?
- Where shall we meet? I would be happy to come to your office if that's more convenient.
- I can recommend a quiet restaurant near the city centre that would be easy for us both to reach.

To be confirmed (TBC)
- I need to check back with my colleague about that. Could I get back to you on that this afternoon?
- Could I confirm that with you tomorrow when I've spoken to my colleague?

It's a date

- Yes, that's fine. I look forward to seeing you at 3.00 p.m. on 25 November at your office.
- That's perfect for me, I can make it then. So let's fix our meeting for 25 November.

> **Avoiding confusion with dates**
> When dates are written in figures in **British English**, the day comes before the month. For example, 9 March 2015 becomes 09/03/15. Note that, in **US English**, the same date is written March 9, 2015 and therefore becomes 03/09/15. Due to these different conventions, it is advisable to write dates out in full.

Useful grammar

The conditional tense, as in "would" and "could", is used frequently in the above phrases. Conditional verbs make questions and suggestions sound open for discussion, rather than fixed and already decided, and therefore lend the suggestion a politer note.

Useful vocabulary

and yet: und doch
to suit: (gut) passen
prospective: künftig
to get back to sb: zurückrufen, sich melden

Rescheduling, cancelling or confirming a meeting

In the interest of informing meeting participants as quickly as possible, it is common to postpone or cancel meetings by telephone or email, especially if this is necessary at short notice. In either case, it is a good idea to give the reason for changing the arrangement if appropriate, and to suggest another time for the meeting to take place.

Rescheduling

Examples

Example 1: calling to reschedule a meeting

A: Hello. This is Sarah from XYZ Com.

B: Oh, hi Sarah.

A: I'm calling regarding our meeting on 6 June at two o'clock. I'm afraid I have to ask if we could reschedule the meeting for the same time on 7 June? I'm very sorry to inconvenience you.

B: That's not a problem. Only, could we make it three o'clock on 7 June instead?

A: Yes, that's fine. Thank you for being so flexible. I look forward to seeing you at three o'clock on the 7th.

B: Okay, see you then. Bye bye!

A: Thanks again. Bye!

Example 2: postponing a meeting by email

Subject: postponement of project meeting of 20 December

Dear Steve

Unfortunately it is necessary for me to change the arrangement we made for next Monday due to the rail strike that has been announced for next week. Please let me know what alternative day and time would be convenient for you.

I apologise for changing our arrangement at such short notice and look forward to hearing from you with regard to an alternative date and time.

Best regards

Simon

Useful phrases

- I'm very sorry, but I'm afraid I have to postpone our meeting of next week, as I've been called to an urgent meeting at our head office. Would it be possible for us to meet the week after instead?

- I'm afraid I've been called away on urgent business next week, which unfortunately means that we have to reschedule the meeting we arranged for next Tuesday. Would any other days next week be convenient for you?

- I was wondering if it would be possible to bring the meeting forward by a week/postpone the meeting until the week after?

Cancellation

Useful phrases

- I'm very sorry, but I'm afraid it is necessary to cancel our meeting of next week until further notice.

- Due to unforeseen complications with the draft contract we are obliged to cancel next week's meeting.

- I very much regret to inform you that we have no other option but to cancel our meeting in Salzburg on Friday.

Confirming a meeting

It is a good idea to confirm an arrangement – especially one made by telephone – in writing. This also provides a good opportunity to give visitors travel directions if necessary.

When you write or email to confirm a meeting, it is recommended to send out travel directions and/or a map for reaching the meeting destination, so as to give your business partner plenty of time to make travel arrangements. See "Hands-on organisation" (p. 22) for tips on giving directions.

Examples

Example 1: confirming a meeting by email
Subject: marketing proposal meeting on 25 November
Dear Mr Smith

Following our telephone conversation of this afternoon, I am pleased to confirm our meeting at our offices on 25 November. The meeting is scheduled to take place from 3.00 p.m. to 4.30 p.m. Please find attached directions to our offices.

I very much look forward to meeting you on 25 November. Please do not hesitate to contact me if you require any further information in the meantime.

Yours sincerely

David Braun

Example 2: confirming a meeting (email to a colleague)
Subject: meeting of 25 March – Peter Smith ok
Wanda

Just to let you know that Peter Smith can make the meeting on 25 Nov after all. I made a reservation for the Arctic meeting room from 3 to 4.30. Hope you can still make it!

Best

David

Useful phrases

As a participant invited to a meeting, you can also confirm by phone:

- Hello. My name is Sarah Hughes and I'm calling from XYZ Com. I am due to attend a meeting at your company on 14 May and I would just like to
  - check the best way to reach your offices by car/public transport?
  - confirm the time of the meeting.
  - check whether there is an overhead projector in the meeting room?
  - find out if there any suitable hotels near to your offices?

Useful vocabulary

at short notice: kurzfristig
to bring it forward: vorverlegen
to decline: (von vornherein) ablehnen
to reschedule: verlegen, neu anberaumen
inconvenience: Unannehmlichkeit
directions: Wegbeschreibung
to postpone a meeting: (nach hinten) verschieben
to attend a meeting: dabei sein, teilnehmen
to make the meeting: schaffen

Making the agenda

All well-structured meetings should have an agenda, which is usually prepared by the chairperson. Depending on the type of meeting, agendas can be formal or informal, but all should start by stating the date, time and location of the meeting.

It is useful to include the name of the person who will be presenting a specific agenda item. You may also find it helpful to include a note of the time allocated to each point. Some more detailed agendas also state objectives for individual agenda items, for example: "Agree on product design".

Formal agendas differ from informal agendas in that they start with routine items, which always appear in a specific order. In addition, each point on the agenda is clearly numbered. Nowadays, agendas for all but the most important company meetings (board meetings, annual general meetings) tend to use an informal style.

Catchwords and abbreviations
Note how both informal and formal agendas have a concise style and tend to be written in note-like form, often omitting articles before nouns and using abbreviations. For example, AOB stands for "any other business", which refers to topics that are not covered by other agenda items or which have arisen after the agenda was distributed.

Examples

Example 1: informal agenda (eg customer or team meetings)

Agenda for end-of-year sales meeting

5 January, 9.00–10.30 a.m., Meeting room 2A

- Presentation of last year's sales figures (Andreas) – 15 mins
- Forecasts and targets for the coming year (Sally) – 15 mins
- Analysis of last year's sales promotions (Peter) – 15 mins
- Proposals for and scheduling of promotions for the current year (all) – 45 mins

Example 2: formal agenda (eg board meetings)

Agenda: Quarterly board meeting

10 April, 10.00–11.30 a.m., Board room suite

1. Apologies
2. Minutes of the last meeting
3. Matters arising from the minutes
4. Presentation of first quarter results (CEO)
5. Departmental presentations (heads)
6. Motions
7. AOB

Compiling the agenda

Besides giving the meeting structure, the purpose of the agenda is to ensure that the time available is only used for discussing the items listed. Some people hold that topics that are important enough to be discussed should be included as items on the agenda, keeping AOB to a minimum. This can be done by asking participants to submit items for inclusion on the agenda. Email is a practical means of doing this, especially when a large group of people is involved. If the agenda is long, or in the case of a formal meeting, you may wish to circulate the draft agenda to participants as an email attachment, rather than including it in the body of an email.

Examples

Example 1: asking for contributions to the agenda
Subject: agenda for end-of-year sales meeting, 5 January

Dear all,

Thank you all for agreeing to attend the end-of-year sales meeting on 5 January, from 9.00–10.30 a.m. in meeting room 2A.

As usual, the items below will be included on the agenda. Please could you let me know by 20 December if you would like any further items to be added to the agenda. I will then circulate the finalised agenda before the Christmas break.

- Presentation of last year's sales figures (Andreas)
- Forecasts and targets for the coming year (Sally)
- Analysis of last year's sales promotions (Peter)
- Proposals for/scheduling of promotions for the current year (all)

Regards,

Simon

Example 2: circulating a draft agenda and asking for input
Subject: draft agenda for quarterly board meeting, 1 April

Dear all,

Thank you for making time for the quarterly board meeting on 1 April, from 10.00–11.30 a.m. in the boardroom suite.

Please find attached a draft agenda for the meeting.

I would be grateful to receive any further submissions for the agenda by 25 March at the latest. Many thanks in advance.

Kind regards,

Simon Webber

Submitting items for the agenda

If you wish to add an item to the agenda, make sure you submit it to the chair before the deadline for contributions. If this isn't possible – eg if an urgent issue has arisen after the deadline for submissions – let the chair know that you would like to include the item under "Any other business". Start your proposal by thanking the chair for the draft agenda and say why you think the item is relevant and should be included.

Useful phrases

- I would like to propose the item "Potential overseas office" to follow item three, as this was discussed at the last managers' meeting and I think it is also relevant to the European sales team.

- I would like to insert the item "Introduction of monthly sales targets" after item two, as this may be the last opportunity to discuss this issue prior to the sales conference next month.

- I recently received important new information from the customer regarding product specifications. If it is not possible to add the item to the agenda at this late stage, I would like to propose including it under AOB.

- I recently came across some software that could be of interest to the team. Since the deadline for submissions to the agenda has passed, I would like to put this item forward for inclusion under AOB.

Circulating the agenda

After you have asked the participants for their contributions to the agenda, it is important to circulate the finalised agenda to them in good time. Email is an expedient way of doing this, but for important external meetings you may choose to send the agenda out by post instead, particularly if you have to enclose other important background documents with the original. This is also a good opportunity to ask participants what technical equipment they require for the meeting and if they have any special dietary requirements.

Example: circulating the agenda

ABC GmbH
Altstr. 26
56710 Altstadt
Germany

Mr P Smith
Development Manager
EFG Co. Ltd
123 New Road
Newtown NEW 1IT
Great Britain

24 July 2010

Agenda for project meeting of 25 August

Dear Mr Smith

Thank you for your contribution to the agenda for the forth-coming project meeting. Please find enclosed a copy of the finalised agenda.

I also enclose a copy of the service agreement, signed by us, to be discussed under item four on the agenda.

I would be grateful if you would let me know by 10 August what technical facilities you will require for your presentation, and also whether you have any special dietary requirements we should note when arranging refreshments and lunch.

We look forward to seeing you on 25 August. In the meantime, please do not hesitate to contact me if you have any questions or require any further information.

Yours sincerely

Simone Roth

Project Manager

Travel and weather tips

If you didn't send out travel directions and/or a map for reaching the meeting destination when you confirmed the meeting ("Confirming a meeting", p. 13), you have another opportunity to do so when circulating the agenda. If your visitors are travelling from countries with different climates, they may also appreciate some tips on the weather and appropriate clothing for the time of year. For example: "Please note that we are approaching the coldest time of year here in Scotland with frequent rainstorms and low temperatures. We recommend bringing warm and waterproof clothing for your visit."

Useful vocabulary

chair (= chairperson): Vorsitzender, Leiter des Meetings
to allocate: zuweisen
motion: Antrag
to compile: zusammenstellen
to circulate: an alle Teilnehmer versenden
item: (Tagesordnungs-)Punkt
prior to: vor
extension of lease: Verlängerung des Pacht-/Mietvertrags

Hands-on organisation

Author Alan Barker *(How to manage meetings)* states: "Ninety per cent of an effective meeting happens before it takes place." Therefore, when organising the practical aspects of a meeting, it is essential to book everything in good time. Think ahead to what the participants might need – technical equipment, refreshments – and ask attendees about their requirements in advance when you confirm the meeting or circulate the agenda.

Giving travel directions

Providing your visitors with clear travel directions can save them considerable time when they come to arrange their trip. A good time to do this is when confirming the meeting or circulating the agenda, to give visitors plenty of time to make their travel arrangements.

Example

You can reach our office/the meeting venue as follows:
By car:
Exit the M25 motorway at Junction 13
Follow the signs for Milton Keynes
At the first roundabout take the second exit to Little Dunbary (follow A25)
Go straight on at the second roundabout
Turn right at the T-junction
Continue straight on until you reach the turning for Little Dunbary Industrial Estate on your right (opposite the Fox & Hounds pub)
You will find us at Unit 98, Road B. Entrance on the left-hand side of the car park.

By public transport:
Nearest train station: Little Dunbary
Bus 42 to Little Dunbary Industrial Estate leaves the bus station in front of the railway station every 30 minutes (on the hour and half hour); journey takes 30 minutes.

From Birmingham airport:
Take the airport shuttle to Birmingham city centre bus station (shuttle leaves the bus stand in front of the airport every 10 minutes); journey takes 40 minutes
Bus 42 to Little Dunbary Industrial Estate (see above); journey takes 15 minutes

On arrival, please report to reception and we will show you to the meeting room.

Giving information on local accommodation

If visitors are travelling from far afield or from abroad, they may need to make an overnight stay close to the airport or to the meeting venue. You can save them much time and help them to arrange things by suggesting suitable places to stay in the area.

Useful phrases

- If you wish to make an overnight stay in the area, we can recommend

  - the Comfortable Inn, which offers reasonably priced accommodation and is located less than 500 metres from our office.

  - the Travellers' Guest House, which is within walking distance of/a short bus/taxi ride from the office.

— the Luxus Hotel, which is a comfortable hotel located in the city centre and easily accessible from the office by taxi.

— the Wings Hotel, which is situated in the immediate vicinity of the airport and accessible by shuttle bus.

Finding out about visitors' special dietary requirements

If people from different cultures are attending the meeting, don't forget to bear their dietary requirements in mind when organising meals or refreshments. Some visitors may have food allergies that must be taken into consideration, so prior to the meeting it is a good idea to ask them to inform you of any special requirements when confirming the meeting or circulating the agenda.

Examples

Example 1: asking about dietary requirements
Subject: project meeting of 25 August

Dear Mr Smith

Please find enclosed a copy of the agenda for the forthcoming project meeting.

As we will be organising lunch for the meeting attendees, I would be grateful if you would let me know by 10 August if you have any special dietary requirements we should note.

We look forward to seeing you on 25 August. In the meantime please do not hesitate to contact me if you have any questions or require any further information.

Yours sincerely

Simone Roth

Example 2: communicating dietary requirements
Subject: project meeting of 25 August

Dear Ms Roth

Thank you for your letter of 24 July enclosing the agenda for the forthcoming project meeting.

Your letter requested information on special dietary requirements. I am allergic to wheat products and would therefore be most grateful if you could make arrangements for gluten-free meals. Many thanks in advance.

I very much look forward to seeing you at the meeting on 25 August.

Yours sincerely

Peter Smith

Useful phrases

- I only eat halal meat/kosher food.
- I am a vegan/vegetarian.
- I don't consume any alcohol – even in cakes and sauces.
- Unfortunately, I'm allergic to gluten/wheat/dairy products/nuts/seafood.
- I have a wheat/dairy/nut/seafood allergy/lactose intolerance.
- I follow a gluten-free/wheat-free/dairy-free diet.
- I'm afraid I can't eat seafood.

Finding out about visitors' technical requirements

Many people bring laptop computers and other devices to meetings and there are a host of other types of equipment they might need you to supply. Don't forget to check in advance what the participants require. If you're responding to the organiser's request for information on technical equipment, make sure you reply before the deadline.

Useful phrases

- I would be grateful if you could let me know by 10 August what technical equipment you will need for the meeting/your presentation.

- Please let me know by 10 August if you require any technical equipment for the meeting/presentation.

- I would be most grateful if you could supply/provide an overhead projector and a local adapter for my laptop power cable.

- I will be bringing my laptop with me to the meeting. I would therefore be most grateful if you arrange for an extension lead to be provided.

Booking meeting facilities

Once you have gathered information on the participants' individual requirements and have a clear picture of what will be needed for the meeting, it's time to book the meeting room, necessary facilities and refreshments.

Examples

Example 1: telephoning to book meeting facilities

A: Hello, I'd like to book a meeting room for the Sales Department from 10.00 to 12.30 on 10 January, please.

B: Sure. How many people will be attending the meeting?

A: There'll be 15 of us.

B: Okay, I'll just check what's available ... Conference Room A is free at that time.

A: That's great. Could you just let me know how many power points are in that room?

B: I'll just check for you ... Room A has six power points.

A: Okay, in that case we'll need two extension leads with five sockets each, as most of the attendees will be bringing laptops.

B: That's fine. I'll reserve two extension leads for you.

A: Thank you. Could I also book an overhead projector?

B: Sure, no problem.

A: And some people will be joining us remotely. Could I book conference call equipment and a technician to set it up for us as well?

B: Certainly.

A: Thanks very much.

Example 2: telephoning to arrange for refreshments

B: Will you be requiring refreshments for the meeting?

A: Yes, please. We'll be taking a break at 11.00 a.m. Please could you arrange for refreshments for 15 people?

B: Certainly. Any special requirements?

A: Oh, yes. One person has a dairy allergy and has requested soya milk.

B: No problem at all. Do you need anything else?

A: The meeting will finish at lunchtime. Could you lay on some sandwiches for around 12.30, please?

B: Certainly. Any particular preferences?

A: Yes. We have two vegetarians and someone who only eats kosher food. Could you make provision for that, please?

B: Absolutely. Do you need anything else?

A: I think that's everything, thank you. If you could perhaps let your colleagues on reception know that some visitors from Japan will be joining us and to show them the way to conference room A?

B: Of course, we'll see to that.

Intercultural considerations

Besides culture-related dietary requirements, other cultural aspects come into play when arranging a meeting involving participants from different parts of the world.

Times of the day and public holidays

Particularly if you are organising a teleconference involving participants in different countries, it is important to ensure that the meeting will take place at a time that is reasonable for all the participants, wherever they are located. Remember, too, to check for important public holidays in your colleagues' countries. The website www.timeanddate.com contains a world clock and a meeting planner comparing times of the day in different countries. It also enables you to create calendars for different countries which display public holidays.

Important time zones		
abbre-viation	explanation	where? when?
Europe		
GMT (UTC)	Greenwich Mean Time (= Coordinated Universal Time)	Stays the same all year round; used all year round in Iceland and during the winter in UK and Ireland
DST	Daylight Saving Time (= Summer Time)	Term used when time is advanced by one hour during the summer time
BST	British Summer Time	UK, summer
IST	Irish Summer Time	Ireland, summer
WET/WEST	Western European Time/Western European Summer Time	
CET/CEST	Central European Time/Central European Summer Time	
EET/EEST	Eastern European Time/Eastern European Summer Time	
Canada and USA		
AST/ADT	Atlantic Standard Time/ Atlantic Daylight Time	"Daylight Time" is used in summer time
EST/EDT	Eastern Standard Time/ Eastern Daylight Time	
CST/CDT	Central Standard Time/ Central Daylight Time	
MST/MDT	Mountain Standard Time/ Mountain Daylight Time	
PST/PDT	Pacific Standard Time/ Pacific Daylight Time	

Important time zones		
abbre-viation	explanation	where? when?
Asia-Pacific		
CST	China Standard Time	Whole of China, all year round
IST	India Standard Time	Whole of India, all year round
JST	Japan Standard Time	Whole of Japan, all year round

The abbreviations "a.m." and "p.m." are often used in English: "a.m." stands for "ante meridiem" (= before noon/midday), "p.m." stands for "post meridiem" (= after noon/midday). "Midday" (12.00 noon) is 12.00 p.m. (and therefore 13.00 = 1.00 p.m.) and "midnight" (12.00 at night) is 12.00 a.m. (01.00 therefore = 1.00 a.m.).

Where confusion could arise, it is a good idea to use the 24-hour clock system for specifying times of the day (in which case you no longer need "a.m." or "p.m."), or to say, for example, "seven o'clock in the morning" (meaning 7.00 a.m.) or "seven o'clock in the evening" (meaning 7.00 p.m.).

Hierarchies and seating arrangements

However, in some countries, hierarchies play a more important role in business culture, so keep this in mind when making seating arrangements.

In general, the seat with the back to the door is the worst, as you cannot see who enters and leaves without turning around. The seats on both sides of the chair signal closeness to the leader. Seating arrangements must be carefully made in meetings with Asians: the top-ranking person should sit as closely as possible to the centre of the table with his or her subordinates on either side, in descending order of responsibility. The hosts should also sit closer to the door to greet the guests.

Asians usually bow as a greeting, with younger people and lower-ranking employees bowing lower to show their respect. An American-style slap on the shoulder would not be the right way of greeting a Japanese business partner and neither would "la bise" (the French way of greeting familiar faces by a slight kiss on both cheeks).

No matter of age
In English-speaking cultures, it is not unusual for junior staff to participate in meetings and for managers higher up the hierarchy to ask their opinions where their particular areas of expertise are concerned. It is acceptable for people at all levels to contribute ideas or ask questions regarding aspects outside their specialist area to gain a clearer overall view of the problem.

Evening entertainment

Culture may also affect your choices of evening entertainment. Whereas dinner is universally accepted as the number one choice for the evening, the Japanese also take their business partners out to a Karaoke bar, and in England it is not unusual to attend a sporting event together or, in informal situations, to have a beer at the local pub.

Useful phrases

Inviting someone out

- Can I invite you to join us for a drink before dinner?
- How about Joe's Karaoke bar for tonight?
- Our company sponsors the local cricket club. Would you like to join us for Saturday's match?

Accepting an invitation

- What a brilliant idea! I'd love to join you.
- Absolutely! I've never been to a cricket match before, thank you!
- I'm afraid I'm tied up all Saturday with a family event.
- I'd rather stay in tonight. I am really tired from the long journey. I hope you understand.
- I'm afraid I'm not really into cricket.

Useful vocabulary

junction: Autobahnausfahrt
roundabout: Kreisverkehr
industrial estate: Industriegebiet
venue: Veranstaltungsort
on the hour: zur vollen Stunde
in good time: rechtzeitig
device: Gerät
overhead projector (= OHP), projector: Beamer [this is a German word, not an English one! In informal British English "beamer" means a BMW]

transparency (for the OHP), slide (PowerPoint): Folie
power point/power socket (for laptops, etc): Steckdose
adapter (for power cables, etc): Adapter
extension lead: Verlängerungskabel
to attend a meeting: besuchen, teilnehmen an
remote: nicht vor Ort, standortfern

Arriving at the meeting

The time before the meeting actually starts is valuable time for socialising and getting to know the other participants better. There is no second chance to make a first impression, as the saying goes, so this is why introducing yourself to others and making introductions for the attendees you know are very important. Last but not least, the physical surroundings and the technical equipment for the meeting also have to be in place and running.

In this chapter, you will find out more about

- welcoming the attendees (page 34),
- introductions (page 38) and small talk (page 40),
- setting up the room (page 45).

Arriving in reception

When arriving at reception, start by giving your name and the name of the person you are here to see.

Useful phrases

- Good morning, I'm here to see Ms Smith. My name is Sylvia Ackermann. I'm a little early, actually.

- Hello, my name is Sylvia Ackermann. I'm here for a meeting with/to meet Hilary Smith.

Being on time

While punctuality is not always strictly observed internally at companies in English-speaking countries, punctuality for external meetings is considered very important. Lateness is seen as impolite, unless there are very good reasons for it. If you are going to be unavoidably late, phone ahead and let the person you are meeting know what is happening – people rarely mind if you have a good reason and keep them informed.

Receiving visitors on arrival

If you are responsible for welcoming visitors, the reception you give them will form their first impression of the company. So it helps to have some friendly phrases at the ready. If your visitors are visiting the company for the first time, they may well have planned in extra time for their journey and arrived a little early. If there will be a short wait while the person they have come to see finishes what they are doing, put them at ease with a little small talk.

Useful phrases

- Hello, Ms Ackermann. I'll let Hilary know you're here. Please take a seat for a moment.

- Good morning, Ms Ackermann. Welcome to ABC Ltd. Please take a seat. Ms Smith will be with you shortly.

- Ms Ackermann – Hilary knows you are here and her meeting is just coming to an end. She will be with you very shortly. Can I offer you something to drink while you wait?

- Ms Smith will be with you in just a moment. Can I bring you a tea or coffee in the meantime? Or some water?

Lift talk

Perhaps you've been asked to show the visitor in reception the way to the meeting room. On the way down the corridor or in the lift, it is customary for both parties to make some small talk to break the ice. In this situation, small talk is generally restricted to topics such as travel, or the weather.

Example: arriving at a meeting and making lift talk

A: Ms Burmeister? Good morning. I'm Rosi Forster, Dr Huber's assistant. Welcome to Hamburg!

B: Thank you, Ms Forster. Nice to meet you!

A: Nice to meet you, too. Did you have a good trip?

B: Yes, thank you.

A: The meeting is on the second floor. Shall we take the lift or would you like to walk?

B: Actually, stretching my legs wouldn't be a bad idea, after so many hours on a plane! But before we go, is there a restroom I could use?

A: Of course, over there behind the column, to your left.

[Five minutes later]

B: So, that's better. Ready to roll.

A: Fine. Let's take the stairs. Is this your first time in Hamburg?

B: Yes, it is. I hope I'll have a chance to do a little sightseeing. I expect the harbour must be especially worth a visit?

A: That's right. It's really interesting. What was the weather like in New York when you left?

B: It was awful, really, slush and snow, and grey skies.

A: It's unusually cold for the time of year here, too. Here we are, room Sao Paolo. Can I take your coat?

B: Thanks.

A: Would you like something to drink? Water? Coffee?

B: Just a sip of water would be nice, thanks.

A: Still or sparkling?

B: Sparkling, please. And would you have a few ice cubes?

A: Sure. I'll be back in a minute.

Useful phrases

- How was your journey here today?

- Did you travel by car? Was the traffic good today?

- Is this your first visit to the area/the company?

- Did you find us easily enough?

- How long will you be staying in Germany?

- I'm afraid you're a little unlucky with the weather. How is the weather in Florida at the moment?

- Well, it looks like we'll have some nice weather during your stay.

If you're the visitor, naturally your responses should be up-beat and cheerful and you should have some questions and phrases of your own up your sleeve.

- The journey was very good, thanks. Ms Smith's directions were very clear.

- Yes, it's my first time here. I'm hoping to see a little of the local area during my stay.

- Has the company been situated here long?

- The company seems to be very conveniently located. Does it take long to travel into the city from here?

- It's much warmer here than where I've travelled from. Is this weather usual for the time of year?

- At home, it's lovely and warm. It's spring-like!

- It's pouring/coming down in torrents/buckets.

- It's absolutely freezing/boiling.

Useful vocabulary

to arrive in reception: an der Rezeption ankommen
restroom (US): Toilette
to stretch my legs: sich die Beine vertreten
to boil: kochen; sehr heiß sein
slush: Schneeregen
a sip: ein Schluck
to cheer up: aufklaren
to be worthwhile: sich lohnen

Introducing oneself and others

When you reach your destination, be sure to thank the person who has shown you the way before they go. The next step will be to introduce yourself to your host and the other meeting attendees. It is usual to shake hands as you do so.

"How do you do?" and "How are you?"

"How do you do?" is a standard phrase people use when meeting for the first time. It is very polite and more formal than "Pleased to meet you." Although this sounds like a question, the required response is: "How do you do?" In the US, people meeting for the first time sometimes substitute "How do you do?" for "How are you?" (response: "How are you?"). This phrase is also becoming more widely used in UK business culture. The phrase "Nice to meet you" also exists and is a little less formal than the alternatives above. "Pleased to meet you" is more appropriate in a business setting. When people who have met on previous occasions ask "How are you?", the usual response is: "Fine, thanks, and you?" or, more formally, "Very well, thank you, and you?". Americans might also reply "Good", which may sound a little informal for speakers of UK English.

Useful phrases

- How do you do? I'm Sylvia Ackermann.
- How do you do, Ms Ackermann? I'm Hilary Smith.

- Hello, I'm Sylvia Ackermann from DEF GmbH in Germany – pleased to meet you.

- Hilary Smith. I'm pleased to meet you, Ms Ackermann.

First name or last name?
It is usual to give both first and last names when introducing oneself or others. However, use of first names is very widespread in English-speaking countries, even between people at different levels of seniority and people who have never met face to face. If they call you by your first name, feel free to do the same. If you use the last name, the usual form of address is Miss (for an unmarried woman or if you are unsure whether she's married or not), Mrs (for a married woman) or Mr.

Introducing others

There are several ways to introduce other people to one another, depending on the formality of the situation. It is useful to follow the introduction with a line about the person you are introducing. The more formal types of introduction are at the top of the following list and the least formal towards the end.

Useful phrases

- Mr Stevenson, may I introduce Peter Korb? Mr Korb is Marketing Director at DEF GmbH in Munich.

- Mr Stevenson, I'd like to introduce Peter Korb. Mr Korb has come from Germany to join us for today's meeting.

- Michael, I'd like you to meet Peter Korb. Peter is Marketing Director at DEF GmbH.

- Michael, have you met Peter Korb from DEF GmbH?

- Michael, this is Peter Korb, Marketing Director at DEF in Germany.
- Michael Stevenson – Peter Korb.
- Michael – Peter. Peter – Michael.

Small talk

As is well known, it is customary for business partners in English-speaking countries to make some small talk before getting down to the business at hand. With someone you have just met for the first time, the aim of small talk is to break the ice and to find out a little about him or her from a business perspective. Generally speaking, the better you know a person, the more topics you can add to the list of small talk subjects. For people who know each other well, the weather tends to be further down the list of small talk topics.

Examples

Example 1: small talk with someone you know well

A: Peter, good to see you again!

B: Same here, Jack! How has life been treating you since we last met? And how is Susanne?

A: Thanks, Susanne is fine. We're going to have twins in July.

B: Congratulations, that's great news.

Example 2: small talk with someone new

A: So you also arrived from London, I heard?

B: Yes, that's right. I flew in from Stansted this morning. Awful traffic on the M11, just one big traffic jam after the other.

A: I know what you're talking about. I live in Cambridge and have to take the M11 every morning to go to work.

B: Poor you! Where exactly are you based?

A: At the Harlow office. Been there for about 18 months now.

B: Do you happen to know Peter Brooks? From accounts.

A: Peter, sure! We used to have adjacent offices. How do you know Peter?

Useful phrases

- Have you travelled far for today's meeting?
- Whereabouts are you/is your company based?
- So, how long have you been with the company?/Have you been at the company long?
- What is your role at the company?

When talking to someone you have met on previous occasions and with whose job and company you are already familiar, you might start by asking the person about how they are. Then look for common ground with regard to other business or work-related topics and industry developments.

- Hi, John. Good to see you. How are you?
- Hello, Mark. How are you settling in in your new role?
- I heard about your promotion to Sales Director. Congratulations!
- Did you attend the technology conference last week? What did you think of ... ?
- I saw your presentation at the annual convention last month – it was really interesting.

- Have you heard that CBC may be merging with Smith Technologies? How would that impact your business?

- Have you been affected by the recent increases in fuel costs at all?

Handshaking

Handshaking is usual during first introductions but not necessarily at every subsequent meeting. While participants in important meetings are likely to shake hands when arriving at the meeting, in informal settings business partners who have met each other on previous occasions and feel relaxed in each other's company may not necessarily shake hands every time they meet.

How's business – and life?

A common phrase English-speakers use with people they have met before and with whom they are on informal terms is simply, "How's business?" This signals your interest in your conversation partner and their company without pushing for any specific information. This phrase leaves it open for your conversation partner to go into as much detail as they wish in their response. Business partners who know each other well may also graduate to small talk topics regarding their private lives.

Useful phrases

- How's the family? Please give/pass on my regards to [name of partner].
- So, how's life treating you?
- Have you been busy of late?
- So, how was your break in Spain?
- Have you managed to get away for a holiday yet?
- Last time we met you were off to the Bahamas. Did you have a good break?

Checklist: levels and topics of small talk

Situation	Example small talk topics
Escorting some-one to a meeting	Journey; travel plans while staying in the area; weather
Post-introductions; meeting someone for the first time	Journey; person's role at their company/their work; information about their company/industry; travel plans while staying in the area; weather
Someone you have met before; waiting for the meeting to start	How are you? How's business?; journey; trade fairs/conferences, etc; industry developments; events of shared interest: work or business-related; weather
Someone you know well	As level above, and also: family; holidays; news, sports, music events, etc

Effortless small talk

To make your small talk more fluent, you should bear some tactics in mind. They will make it easier for your partner to carry on with the small talk, and you will be perceived as a person who takes an interest in others and is open and friendly:

Checklist: tactics for effortless small talk

Tactic	Example
Give answers in full sentences.	Where in Germany are you from? I'm from Passau, that's in the south, close to the Austrian border.
Listen for indirect signals.	We had quite a storm on the lake yesterday and almost capsized. So you must be a good sailor?
Give some information about yourself.	I really need to leave the meeting on time. I've got my ballet class at seven.

Useful vocabulary

to merge, a merger: fusionieren, Fusion
to impact sth: sich auswirken auf
to be affected by sth: betroffen sein von
adjacent: nebeneinander liegend

Setting up the meeting room

Even if you reserved the necessary technical items, things don't always run smoothly on the day of the meeting itself. You may need help setting up the equipment – especially if you are using conference call equipment. Ask a technician to check if it is set up properly before the meeting starts.

Useful phrases

Asking for help

- I'm afraid this extension lead isn't quite long enough. Could I trouble you for a longer one?
- I'm having a little trouble with the projector. Please could you give me a hand to set it up?
- Is there anywhere I could plug my mobile phone in to charge during the meeting?
- Do you mind if I plug my laptop into this power socket?

Offering help

- Are you okay with that? Do you need any help at all?
- Can I give you a hand with setting that up?
- Do you need an extension lead for your laptop?
- Just a moment – I'll ask someone from technical support to come and give you a hand with that.

Useful vocabulary

to give/pass on regards to sb: Grüße ausrichten, grüßen lassen

to give/lend sb a hand: helfen

Conducting a meeting

Besides playing a key role in preparing the agenda, the chairperson gives structure to the meeting itself. He or she provides a framework for discussion of the agenda items by performing the formalities at the start, throughout and at the end of the meeting.

In this chapter, you will find more information on

- opening the meeting (page 48),
- guiding the discussion (page 51),
- bringing about a decision (page 56),
- closing the meeting (page 57).

Opening the meeting

The main task of the chair is to make sure the meeting runs smoothly. He or she opens and closes the meeting, manages time, facilitates conversation, assigns tasks and summarises what was achieved during the meeting.

Example: the start of the meeting

Chair: Good afternoon, everyone. Before we start, Susanne Braun has asked me to pass on her apologies: unfortunately she can't make it today as she's attending the sales conference in Athens. I would also like to introduce David Barnes from ABC Europe who has flown in from Brussels to join us today. David, could I ask you to tell us briefly about your role in the project?

DB: Yes, thank you, Stephan. I'm heading up the project team in Brussels. We're a team of six and we're dealing with the logistics side of the product launch in Europe.

Chair: Great, thank you, David. For David's benefit, could we all introduce ourselves, going anti-clockwise around the table?

[Round of introductions]

Chair: As you all know from the agenda, today's objective is to agree on the packaging for the new product. Now, we do need to make sure we finish on time this afternoon so that David can make his evening flight back to Brussels. We've allowed 15 minutes for each item on the agenda and a further 20 minutes for questions at the end. So, shall we get straight on with item one?

Useful phrases

- Good morning, ladies and gentlemen. Thank you all for joining us today. Shall we make a start? [Formal]

- Good morning, everyone. Thanks for coming. Shall we get down to business/get the ball rolling? [Informal]

Introductions and apologies

At external meetings where the participants do not yet know each other, the chair is likely to start by making formal introductions or asking the participants to introduce themselves.

Useful phrases

- Perhaps we could start by introducing ourselves, moving clockwise around the table?
- For those of you who haven't yet had the chance to meet, I would like to start by introducing ...
- Our presenter today is Sarah Hughes. Sarah is going to talk to us about ...
- On my left is ...
- Also joining us is ...
- I have received apologies from David Smith, who is unable to join us today, and Fred Burnes, who will be arriving a little late.
- Has anyone seen Martha? She's supposed to come, too.

How to say "I'm sorry"

Whether to use "apology" or "excuse" depends on the situation. In a meeting, you use "apologies from Paul", meaning that Paul was unable to attend. An "excuse" can also be used in the sense of the German "Ausrede, Vorwand". In formal English you would say, "I apologise for being late", in informal English "I'm sorry I'm late". In both cases, you would add the reason for being late, eg: "My plane was delayed".

You use "excuse me" to catch someone's attention, eg: "Excuse me, could you pass me the coffee, please?" "I beg your pardon?" or "Pardon?" are formal ways of stating that you haven't understood what someone was saying. In informal situations you simply say "Sorry?".

Introducing the agenda and the objectives of the meeting

Useful phrases

- As you know, we're here today to discuss ...
- The objective/purpose of today's meeting is to reach agreement on ...
- Does everyone have/does anyone need a copy of the agenda?
- We have approximately 15 minutes for each point on the agenda.
- So, moving onto the first item/point on the agenda, ...

Initiating the discussion

Useful phrases

- The first agenda item – renewal of company insurance – was put forward by Jill Baker. Jill, would you like to start us off?
- Simon, you put forward item three on the agenda. Perhaps you could fill us in on this point?
- Jürgen, could you bring us up to date on item four?

- Stephan, would you outline item five for us?
- Peter, could you fill us in on the background?

Useful vocabulary

to get down to business, to get/start the ball rolling: beginnen, anfangen

clockwise, anti-clockwise: im Uhrzeigersinn, gegen den Uhrzeigersinn

to fill sb in on sth: informieren

Guiding the discussion

It is the chair's responsibility to make sure all the participants have their say – by guiding the discussion, encouraging everyone to speak up and, if necessary, curtailing the flow of speech of individuals who are too vociferous.

Useful phrases

- Thank you, Stephan. Peter, do you have any thoughts on this?
- So, Peter agrees with Stephan regarding this item. Barbara, could I bring you in at this point?
- I'd also like to hear Julie's opinion at this point.
- Could you elaborate on that for us, please?
- Does anyone else have anything to add to this?

Dealing with dominant participants and interruptions

Useful phrases

- Thank you, Jim. Could I just stop you there for a moment, because I'd like to bring Phil in here.

- Jim, could I ask you to let Phil finish his point and then we can hear your view?

- I'm sorry, Jim – I'd just like to finish hearing what Phil has to say on this point and then we'll come back to you.

- I'm sorry, Tatjana, we don't have time to talk about that now, I'm afraid.

> **Tricks for the chair**
> Notice how the chair phrases requests as suggestions or questions. He addresses the participants by name to secure their attention and handles interruptions tactfully with "I'm sorry", followed by an assurance that the person concerned will have their say afterwards.

Encouraging quiet participants to contribute

Example: facilitating a balanced discussion

Chair: Item one is the font used on the outer packaging. Jo, could you outline the design department's view on this?

Jo: Sure. We've come to the conclusion that Option A is the best choice as it's the most legible against the blue background and consistent ...

David: Can I come in here? I think ...

> Chair: Sorry, David, could we just let Jo finish her point and then we'll hear your view.
>
> Jo: ... and it's consistent with the overall branding.
>
> Chair: Thank you, Jo. Did you want to comment on that, David?
>
> David: Yes – in Logistics we strongly prefer Option B. It's far more eye-catching.
>
> Chair: So we have Option A for Design and Option B for Logistics. Sue, could you please tell us Manufacturing's perspective?

Useful phrases

- I'd like to ask Petra to tell us about the new development at this point, if I may.
- Mr Baker, what is your opinion here?
- Jim, it would be useful to hear your perspective on this.
- Pam, how would this affect the HR [Human Resources] department?
- Julie raised the point that our existing contract is about to expire. What's your take on this, Jan?
- Tara, you've been working with this client for some time now. What do you say to the point Helen raised?

Reminding participants to be brief

Useful phrases

- Before we move on to the next item, could you quickly give us your opinion on this, Maria?
- I'm afraid we have to move on to the next item. Could you bring your point to a close, please?

- I'm afraid I have to stop you there. Time is ticking away.

- Could I ask you to keep your comment brief, please?

- Can I remind you that we only have 15 minutes for each item?

Keeping to the agenda

Example: directing the discussion

David: ... and therefore we would be strongly in favour of B, because, as I've already said, we ...

Chair: Could I stop you there for a moment, David? I'm afraid time's running short. Jo, what do you think about item two?

Jo: Design would like to rethink the shape of the packaging. We'd like to make it a little taller, because ...

David: I totally agree. Some years ago we opted for square packaging for another product and it was a total flop. In fact, this particular product was ...

Chair: That's a valid point, David, but I'd like to bring the discussion back to this year's launch if possible ...

Useful phrases

- I think we risk moving away from the agenda here. Could we bring the discussion back to ...?

- We have a lot to cover this morning – I'm afraid we don't have time to go into that. Can we focus on ...?

- I'm afraid we're getting sidetracked. Let's return to ...

- That's an important point, but I think that's a discussion for another day. I'd like to go on with the sales figures.

- That's a valid point. May we come back to that later?

Summarising and concluding an item

Useful phrases

- So, we've established that the July deadline is too ambitious. What is a more realistic target?
- So, just in order to summarise/sum up/recap what we've said, ...
- So, to bring this point to a close, we can say that ...
- So, am I right in concluding that we'll ... ?
- I think we've agreed/we're all in agreement that ...

Moving on to the next agenda item

Useful phrases

- The next point/item two on today's agenda is ...
- I'd like to move on to item three now.
- Now, turning to item four, ...
- Anyway, about the new location ...

Useful vocabulary

to raise a point: einen Punkt ansprechen
to expire: auslaufen
font: Schriftart
to focus on sth: sich konzentrieren auf
square: rechteckig
to elaborate: etwas ausführen
ambitious: ehrgeizig

Bringing about a decision

The purpose of most meetings is not only to exchange information, but also to reach decisions. If a decision does not come about naturally, the chair might have to "push" the participants to make up their minds. Sometimes it may also be necessary to postpone taking a decision.

How to reach consensus

Attendees need sufficient time to express their opinions. Nobody likes to be rushed into a decision. Minority views should receive ample attention so that their owners feel they were heard. A short break often helps to bring about a decision.

Summarising the general mood at the end of each item on the agenda contributes to a clear outcome. Every attendee is more aware of what the general views are. The chair can also help the participants reach a decision by phrasing questions carefully. If the chair has the feeling that there are more outspoken supporters in favour of an issue, they should ask: "Does anyone object?" rather than "Does everyone approve?"

Example: summarising and bringing about a conclusion

 Chair: So, to sum up item five, we've agreed that we're going to opt for silver packaging. The next item on the agenda is "Cardboard or plastic packaging". We've already heard everyone's views on this topic, so could we move straight to a show of hands for cardboard? That's six. Thank you. For the minutes, that's a majority decision in favour of cardboard.

Useful phrases

- Could we make a yes or no decision on this item? Could all those in favour please raise their hands?
- Could we have a show of hands for going ahead with the measure, please?
- Am I right in saying that we've decided in favour of/ against this move?
- So we have decided to commission Union Brozers with the catering for the event.
- Is it okay with everybody to bring the launch forward by two weeks?
- It would certainly be wrong to rush into a decision. Could we leave this until another time?

Useful vocabulary

to opt for: sich entscheiden für
majority decision: Mehrheitsentscheidung
show of hands: per Handzeichen

Closing the meeting

Initiating further action

Before the meeting closes, all the "to dos" should be allocated. In the closing remarks, the chairperson or participants may also wish to discuss the date and time for the next meeting, when the minutes will be available or the date by which a decision should be made.

Example: allocating follow-up tasks

Chair: Jo has kindly offered to follow this up with the shipping department. Can you do this before next week's meeting, Jo?

Jo: Yes, no problem.

Chair: Great. Can somebody volunteer to contact the supplier before next week? David? Thank you. We're almost out of time, but does anyone have any last questions before we finish? No? Well, so we can finish the meeting. Thank you all for coming and for your input. I wish David a safe flight back to Brussels.

Useful phrases

- We need somebody to contact the supplier. Ian, could you do this for us?

- Tina has offered to file the application. How long do you think this process will take, Tina?

- Could you follow this up for us, please?

- Thanks for volunteering to do this, Marina. Do you think this will be possible in time for next week's meeting?

- Georg, when will you be able to email out today's minutes to everyone?

- Betty, do you think you could get back to us on that to-morrow by email?

- We'll meet again on the first of next month.

- I'll send out a group email with the minutes tomorrow.

- Can we fix the date for the next meeting, please?

- So, the next meeting will be on next Tuesday.

- What about the following Wednesday? How is that for everyone?

Bringing the meeting to a close

There are different reasons why a meeting comes to an end. Time may have run out or all of the items on the agenda may have been checked off. Some meetings will end earlier than expected and others will run late. Before the chair closes the meeting, he or she will let the participants know the meeting is drawing to a close.

Useful phrases

Formal meetings

- I am officially ending today's meeting. You will receive the minutes within two days.
- I declare the meeting closed.
- The meeting is adjourned until tomorrow, 8.00 a.m.

Informal meetings

- It looks like we've run out of time.
- Before we bring the meeting to a close, does anyone have any other business to discuss?
- Well, we're almost out of time for today. Are there any last questions before we finish?
- Is there any other business?
- I think we've covered everything on the agenda.
- As you can all see from the agenda, that was the last item. If no one has anything else to add, then I think we'll wrap this up.
- That brings us to the end of the meeting.

- Let's call it a day, then.
- The meeting is closed.

Thanking the attendees

Thanking the participants is usually one of the last remarks. However, even after closing the meeting the chairperson might realise they have forgotten something. There is almost always one last thing to say.

Useful phrases

- I'd like to thank you all for coming and I wish you a safe journey back.
- Thank you all for your participation.
- I'd especially like to thank Karl for coming over from Glasgow.
- Oh, before you leave, please make sure you have signed the attendance sheet.
- Could I have your attention again, please? I didn't mention that …
- If you could all return your chairs to the room next door that would be much appreciated.

Useful vocabulary

to adjourn a meeting: vertagen
to call it a day: beenden, Schluss machen
to wrap sth up: abschließen
to follow sth up: an etwas dranbleiben

The meeting itself

There are different roles and goals in a meeting. The chair is in charge of smooth and efficient "housekeeping". However, participating also has its pitfalls and requires particular types of language for achieving one's aims, above all when it comes to politeness and expressing oneself diplomatically.

On the following pages you will find information on

- distributing and accepting roles (page 62),
- active listening and asking questions (page 64),
- expressing agreement and disagreement (page 69),
- making suggestions and giving your opinion (page 78),
- enquiring and resolving misunderstandings (page 81),
- diplomacy and politeness (page 85),
- what to do in case of language problems (page 90),
- voting (page 91).

Roles at a meeting

Whether your meeting is formal or informal, it will run more smoothly if one of the participants assumes the role of chair ("Conducting a meeting", p. 47).

Other important players at the meeting are the minute-taker and the participants. The minute-taker writes the minutes, i.e. he or she records the meeting and keeps track of what has been said. One could also say that the minute-taker is the administrator for the meeting.

The participants can take various roles, such as the role of the presenter, who provides information, or the role of a task owner, meaning that an individual is in charge of a certain topic. Participants are also thinking resources – they contribute ideas, help to solve problems and shed light on issues from different perspectives.

Assigning and accepting roles

For formal meetings, the above roles will usually be allocated before the meeting itself. At more informal meetings, one of the chair's first tasks during the meeting may be to assign these roles to other participants.

If this role has not been assigned yet – which usually happens in the run-up to the meeting, as the minute-taker should prepare in advance (writing equipment, preliminary information, etc) – perhaps the easiest way is to ask for volunteers.

Example: asking for a minute-taker

Chair: Do I have any volunteers for minute-taker today? Don't all jump at once! No volunteers? Okay. Peter, could I ask you to take the minutes today?

Peter: I'm afraid I'm pretty tied up this week – I don't think I'd be able to turn them around in time for the next meeting.

Chair: No problem. How about Sarah?

Sarah: I'm sorry, but I'm on a training course in the Manchester office all next week.

Chair: I see. Jack, do you have time to do the minutes?

Jack: Yes, certainly.

In the above example, two participants use the following polite formulations to decline the chair's request: "I'm afraid", "I'm sorry, but ..." Notice how the chair continues to ask around the group until this important role has been assigned.

Useful phrases

- Do I have any volunteers for taking the minutes today?
- Who would like to be our minute-taker this morning?
- Bill, could I ask you to be minute-taker today?
- Jane, would you be so kind as to take the minutes for us?

Useful vocabulary

minute-taker: Protokollführer

to assume (informal: to take on) the role of: Rolle übernehmen

to be tied up with sth: viel zu tun haben

to turn sth around: etwas fertig machen und zurücksenden

Active participation and asking for more information

While it is the chair's job to manage the meeting, meeting participants also have an active role to play to aid communication and make the meeting as effective as possible. Asking questions is an effective method of obtaining more in-depth information, while active listening shows the person you are talking to that you are paying attention and have understood what they are saying.

Interrupting politely

Sometimes you may need to interrupt a speaker in order to ask your question before they move on to another point. Perhaps the speaker has finished their point and there is a natural pause in which you can ask your question.

Sorry!
This is the easiest and fastest way of attracting attention politely and stopping the speaker. This "catchword" should, however, be backed up by saying what you actually want. If you just want to check back if you have correctly understood what the speaker said, you could say: "Sorry – how much did you say?" or, "Sorry, when did you say?"

Useful phrases
Formal meetings
- Sorry, I'd just like to ask a question, if I may?
- Sorry, could I interrupt you for a moment? I'd like to ask a question.

- Excuse me, Peter, I wonder if I could interrupt you for a second?
- Sorry, could I come in here with a question?

Informal meetings

- Sorry to interrupt, but I have a question.

If, on the other hand, you are the speaker, you can prevent an interruption by either just ignoring the person or by saying one of the following phrases:

- Sorry, John, can you hear me out, please?
- Just let me finish, please.
- No, Mary, please hear me out.

Asking for more information

Once you have politely gained the speaker's attention, move straight on to your question. Asking for a person's opinion when you ask them a question – for example, "What do you think will happen?" rather than "What will happen?" – helps "soften" the question and makes it less direct. Notice, too, how the questioner below uses "would" and "could" instead of the more direct "will" and "can".

Example: asking for and receiving further information

A: Moving into the spring, we need to reassess the product range. Then, in the summer, we'll turn our attention to ...

B: Sorry, could I interrupt you there for a second? A: Sure.

B: I have a question: you mentioned reassessing the product range in the spring. Could I ask you to expand on that?

A: No problem. Sales of C123 and C124 dropped dramatically last year, so we want to look at the reasons for this and perhaps make some adjustments to the range.

B: Do you mean introducing new products?

A: Actually, it's more likely that some products will be discontinued.

B: I see. What timescale do you have in mind for this process?

A: Well, we're planning to start the assessment in the first week of March and we anticipate that we'll be finished by mid-April.

B: Right. That's good to know – thank you.

Useful phrases

- You mentioned that [repeat the key point]. Could you expand on that for us?

- Could you elaborate/go into more detail on that for us?

- Could you explain that in a little more detail?

- I'm afraid I don't quite follow you there. Could you be a little more precise/specific?

- How do you think that will affect ...?

- What do you think would be the outcome of that?

- What do you consider to be the highest priority here?

- How do you envisage implementing ...?

- What is the timescale likely to be for that/what is the anticipated timescale for this?

- Do you foresee any problems/issues/difficulties there?

- What do you think would be the possible repercussions of that?

Active listening

If you don't want to interrupt the speaker's flow, active listening can simply take the form of single words or "polite noises", such as "okay", "aha", "oh", "mmm". Body language, such as nodding and smiling from time to time, should of course accompany all your polite utterings to show that you are listening. Use the handy phrases below if you want to signal to the person speaking that you understand.

Useful phrases

- I see.
- I understand.
- Right.
- That's interesting.
- Oh, really?
- That makes sense.
- Okay, thank you for explaining that.

Responding to questions

If you need a couple of seconds to think before answering a question, one trick is to repeat the asker's question and clarify what they want to know while you gather your thoughts. When a speaker needs more time to think, they may also try to ask a question back. Maybe they don't have a good answer up their sleeve right away. In this case, they can simply acknowledge the question and play for time before giving a real answer. English-speakers frequently also introduce their answers by starting with, "Well …".

Example: buying time to think and respond

> A: So, how do you see this measure taking shape?
>
> B: How do we see the measure taking shape? Well, first of all, we plan to …

Useful phrases

- I'm not quite sure what you mean by that. Could you explain?
- It all depends on what you mean by the "extra costs" you mentioned.
- That's a very good question.
- I'm glad you asked that.
- It's hard to say.

Or, perhaps you've been caught on the back foot by an awkward question you weren't quite expecting and need to politely decline to answer.

- I'm afraid I'm not in a position to/able to comment on that/answer that question just now.
- Well, it's rather difficult to say at present.
- I don't have enough information at my disposal to consider all the implications at the moment.
- Maybe we could leave the legal issues aside for a moment, the real challenge for ACME is on the European level.

Useful vocabulary

timescale: Zeitrahmen

to envisage: voraussehen, ins Auge fassen

to anticipate: voraussehen, rechnen mit
specific: besondere, speziell, präzise
at sb's disposal: zur Verfügung haben
outcome: Ausgang
to foresee: vorhersehen, absehen
repercussion: Auswirkung

Expressing agreement and disagreement

In English-speaking cultures, politeness and tact are key elements when expressing approval and disapproval of other people's suggestions and ideas. When expressing disagreement – the more sensitive of the two areas – many English-speakers, and particularly the British, tend towards understatement and often use diplomacy so as not to sound impolite.

Agreeing with an opinion

Example: expressing agreement

A: So, I really think we should wait until next year to launch the campaign.

B: I couldn't agree more. I can't see how launching a new model this late in the season will benefit the range.

C: You're quite right – we could put the remainder of this year's budget to far better use.

A: Well, I'm glad we're all agreed on that.

Cultural differences between the US and Britain
Although politeness is important in both the US and British cultures, North Americans tend to be more forthcoming and direct than their British counterparts when expressing their agreement, and especially their disagreement.

Useful phrases

Total agreement
- I couldn't agree with you more.
- You're absolutely right (there).
- I totally/completely/fully/absolutely agree (with you on that point).
- I'm in total agreement with that.

Neutral agreement
- I agree.
- I would agree with that.
- I'm with you on that.
- I think you're right.
- That's a fair point.
- That's true.

Mild agreement
- I tend to agree.
- Maybe you're right.
- I suppose so.
- Possibly.
- Could be.

Diplomatic disagreement

In English-speaking cultures, disagreeing without putting forward an alternative solution is generally viewed as unhelpful. Therefore, the emphasis is on what is known as constructive criticism, which essentially involves coupling disagreement with an alternative suggestion and backing it up with good reasons. So, people often handle disagreement by highlighting an element of an idea they find positive before moving on to tactfully express disagreement with another aspect.

Example: disagreeing diplomatically

A: I hear what you're saying about that but I do have some reservations about the timing.

B: I see. Could you be more specific?

A: Well, we have the international sales conference coming up in September. Surely we should aim to finish the proposal by then so that we can discuss it at the conference.

B: That's a good point, but I think rushing the proposal to finish it before the conference could be counterproductive.

A: I'm just a little concerned that the French sales team will be left out of the discussion completely if the issue isn't addressed at the conference.

B: I take your point, but I think we'll struggle to compile all the data we need by September.

A: Oh, I see. In that case, what about putting together an overview of the proposal in time for the conference and then scheduling a meeting to discuss it in full at a later date?

B: Yes, that would work.

Useful phrases

- I like what you said about the book launch, but I feel we might need to rethink some aspects of it.

- I think the concept is good overall, but I'm not entirely convinced about the marketing element.

- I'm concerned that advertising in this way won't achieve our objective, because ... Have you thought about the possibility of ...?

- My only concern with that is that we could run out of time. Have you considered trying ...?

When "yes" means "no"

The German culture is what is known as a low-context culture. In such cultures, it is considered most effective to formulate statements in an unambiguous and direct way. Americans are also known for their no-nonsense, straight-to-the-point approach. In high-context cultures, such as in Britain, non-verbal communication, including body language, plays a more important role. High-context culture speakers are often reluctant to say "no" in a straightforward way, so be sensitive to "yes, but" statements that really mean "no":

- Yes, I take your point, but ...

- Yes, I agree with you, but ...

- Yes, I see what you're saying/what you mean, but ...

When you want to answer in the affirmative way, there are many possibilities besides a simple "yes". The same applies to "no" and "maybe", as you can see from the following table.

Alternatives to the words "yes", "maybe" and "no"	
Yes	Sure
	certainly
	of course
	right
	hmm
	yeah
	okay
	fine
	I am/I was/I did
	I think so
	absolutely
Maybe	Perhaps
	could be
	I don't know
	it's hard to say
	I'm not sure
No	Not really
	I don't think so
	not just now/not at the moment
	I'm not/I wasn't/I didn't
	not completely

Degrees of disagreement

Bear in mind that, out of politeness, particularly British speakers tend towards understatement in their expression of disagreement. Therefore, phrases that express mild disagreement often conceal opinions that are stronger than they sound. If you're the speaker, remember that toning down your statements a touch will still have the required effect.

Example: various grades of disagreement

A: Yes, I agree, but that target is fairly unrealistic based on last year's figures.

B: I'm not sure I agree there. As we said, last year's turnover was down due to external factors beyond our control.

A: Yes, but that's not to say that the targets weren't too high in the first place.

B: I beg to differ. We were on track to hit our targets before the general downturn in the industry.

A: I see what you're saying, but surely our targets should have allowed some leeway for that possibility.

C: I totally disagree. Sales have increased year on year for the past five years ...

Checklist: disagreement

▪ Politeness and diplomacy are key.

▪ Bear understatement in mind, both when interpreting other people's statements of disagreement and formulating your own. Making your statements milder will still get your point across in the English-speaking world.

▪ If possible, highlight any items you agree with before being specific about the aspects you disagree with.

▪ Use constructive criticism: always try to put forward alternative solutions to problems.

▪ Back up disagreement with good reasons.

▪ "Yes, but" statements are a tactful way to say "no".

Useful phrases

Mild disagreement
- I'm not sure about that.
- I'm not sure I agree.
- I'm not totally convinced about that.
- I hesitate to agree with you there, because ...

Neutral disagreement
- I'm sorry, but I can't agree on that point.
- I'm afraid I don't agree.
- I tend to disagree.
- I beg to differ.

Strong disagreement
If a speaker strongly disagrees with a standpoint, "softeners", such as "I'm afraid", "I'm sorry, but" tend to fall away:
- I totally disagree.
- I don't agree at all.
- I'm in complete disagreement with you.
- To the contrary, I think ...

Expressing criticism

As with disagreement, diplomacy is called for when it comes to expressing criticism in the English-speaking world. English-speakers often tone down criticism by using a positive word in the negative, rather than a negative word:
- not very good: bad
- not up to standard: below standard
- not very encouraging: disappointing.

Try to soften criticism
Criticism is usually received best if it is expressed diplomatically. To soften a statement in English, try using:
"not quite/not really" + a positive word (eg "adequate") or
"a little/somewhat" + a negative word (eg "disappointing").

Criticism: can you read between the lines?	
What is said	**What is meant**
This isn't quite what we were expecting. This falls a little below our expectations.	This doesn't meet our expectations at all.
The design isn't quite what we were hoping for. The design is not exactly what we were looking for.	We don't like it at all.
Their performance isn't really up to scratch. Their performance is a little under par.	Their performance is not acceptable.

Straight talking

Despite your best efforts to put your point across tactfully, sometimes there are situations which call for a more direct approach. In English, unpleasant news is often preceded by short phrases which brace the listener for what is to come.

Example: let's face facts

A: I accept that sales were a little lower than expected.
B: Let's not beat around the bush: they were very disappointing.
A: Frankly, I think that's overstating it a little. Those kinds of figures are to be expected in the current business climate.

B: Look, I think we should face facts: the product is just not performing as we planned. To be honest, I think it's time we pulled the plug ...

Useful phrases

- Let's not beat around the bush ...
- Let's face facts ...
- Let's be honest ...
- Let's get one thing straight ...
- To be honest, I ...
- Frankly, I ...
- I don't want to rock the boat/upset the applecart, but ...
- I don't want to paint too black a picture, but ...

Useful vocabulary

remainder: Rest
range: Produktpalette
to conceal: verstecken
to brace: wappnen
to beat about/around the bush: um den heißen Brei herumreden
current: aktuell
to rock the boat/upset the applecart: die Pferde scheu machen

Making suggestions and having your say

Example: making suggestions

> A: I would advise the Board to slash the budget for tobacco advertising. That way, we are on the safe side.
>
> B: Why do you think that?
>
> A: The new EU directive poses an immediate threat to our market share.
>
> B: In my view this prospective danger is still in the stars. What I think is that we should deal with the problems at hand, rather than musing about what might happen in ten years' time!
>
> A: I'm not dodging the issue of teenage smoking, if this is what you mean. On the contrary! We'll have to face the music sooner or later. I assure you that I do care about the issue.

Informal formulations and questions for making suggestions are also suitable for meetings. A more tentative way of making a suggestion includes "would", "could" or "might/may".

Useful phrases

- Why don't we try ...?
- Let's take a novel approach in this matter.
- Shall we ...?
- May/might I suggest we stick to our initial plan?
- Wouldn't it be worth trying to give a new supplier a chance?
- Could there be another way forward?

Expressing your opinion

Of course, you can simply state your opinion by plainly saying what you mean. You would usually use an introductory phrase like "I suggest" or "I would advise that…"

In a discussion, your tone of voice and intonation would of course also influence the way your opinions come across. Quite naturally, the point you want to make should be pronounced more emphatically than the rest of the sentence. Another way of emphasising and adding stress is to use auxiliaries that you would not need for grammatical reasons.

Useful phrases

Neutral expression of an opinion

- I suggest that we reconsider our standpoint.
- My proposal is to start negotiations as fast as we can.

Strong expression of an opinion

- I strongly recommend investing in commodities.
- It is high time we became active in this field.
- I advise all of you to reconsider your views.
- There's no alternative to buying new machines.
- It's not a rumour, our competitors are launching the new storage system, that's a fact.
- I do think it's important to meet before the 17th!

However, the best phrases won't make an impact unless your entire performance radiates confidence. Here are some tips.

Checklist: how to have your say

- Try to avoid a thin voice, hurrying or stuttering – there's no need to be afraid.

- There's no need to hesitate if you know your point is valid. If you are not sure yourself, say that you are not sure.

- Look at the group: make eye contact with everyone in the group, don't avoid it by looking at the ceiling, the flip-chart, the walls ...

- Play with your voice: emphasise important points, breathe deeply to make your voice stronger, make your voice more interesting by modulating your pitch.

- Express yourself accurately: no generalisations, avoid words like "never", "always", "everyone".

- Keep your hands and feet still – no fiddling or fidgeting.

Useful grammar

Don't forget that question tags may also be used to make a statement more palatable to your counterpart and to urge him or her to react. Note that a negative tag asks for agreement, whereas a positive tag shows that you are looking to your counterpart to share your disagreement.

- I think April is a good month for team trainings, don't you?
- Mr Meyer isn't the right person for the job, is he?

Useful vocabulary

to slash: radikal kürzen

to pose a threat to: eine Gefahr darstellen für

directive: Richtlinie

to fiddle: herumfummeln

to fidget: auf dem Stuhl herumrutschen

to face the music: die Suppe auslöffeln

emphatically: betont

Enquiring and resolving misunderstandings

Remember, it is always better to resolve a misunderstanding immediately so that it cannot lead to problems later or cause the meeting to drift off in a completely different direction. One way to check the facts is to ask for them again.

Asking for repetition

Example: sorry?

A: So, overall, the figures for ... [sound of loud drilling outside]

B: I'm sorry, it's terribly noisy with the window open. I didn't hear your last point. Could you run through it again, please?

A: [speaks very quickly] I was just saying that until we look at the figures we just aren't in a position to know if it actually made a difference to the bottom line and it would be premature to move onto A3 without taking this step. We really should hold fire on this for a while.

B: Okay, so just to clarify, by A3 you're referring to the summer promotion?

A: That's right.

> B: I'm not sure I fully understand the point you were making about A3.
>
> A: In a nutshell, we need to gauge the success of the previous promotion before launching the new one.
>
> B: I see – and yes, I totally agree.

A good way to ask your counterpart for repetition or clarification of what they said is to use the following question words:

- *What* did you say?

- *Who* did you say?

- *When* did you say?

- *Why* did you say?

- *How many/much* did you say?

Useful phrases

- I'm sorry, I didn't quite catch that/the last thing you said.

- I'm sorry, it's a little noisy outside. I didn't hear what you just said.

- I'm sorry, could you speak a little louder, please?

- I'm sorry, would you mind repeating that last point, please?

- I'm sorry, could you repeat your first point for me, please?

- Sorry, could you explain that again, please?

- I'm afraid I'm not quite clear what you mean.

- Sorry, I don't quite follow you.

- Sorry, I'm not sure I understood. Would you mind going over that again?

Summarising for clarification

Another good way to check your understanding is to reformulate the speaker's point in your own words.

Useful phrases

- Are you saying that you can't deliver next week?
- Correct me if I'm wrong, but ...
- Did I get it right that ...?
- Have I understood you right? Do you mean ...?
- So, just to check I understand you correctly, do you mean ...?
- Just to clarify, do you mean ...?
- So, just to make sure we're all singing from the same hymn sheet, you're saying that ...
- To quickly summarise, your point is that ...
- Just to recap, are you saying that ...?

Asking double questions
In natural English, speakers often double up a question to check what somebody said or to find out what they meant, for example:
Warranty? What exactly do you mean by warranty?
Whereabouts is Kemer? I mean, how far is it to the nearest airport?

Recapping and confirming

If you're the speaker, you may feel it is important to clarify your point or to check that the other participants have understood.

Useful phrases

- What I meant was ...
- What I mean to say is that ...
- The point I'm making is ...
- Are you with me so far?
- Are you following me?
- Does that make sense?
- Do you understand what I mean?
- I'm afraid that isn't quite what I meant.
- There seems to have been a slight misunderstanding.
- Maybe I didn't make myself clear.
- The point I'm making is that we are wasting a lot of man-power with the old equipment.
- Actually, I am not talking about renting, I am talking about profitability.

Useful grammar

There are specific verbs in the English language that must be followed by a gerund (-ing form). "To mind" is such a verb. Other verbs and fixed expressions of that kind which are useful for meetings are as follows:

- This would *involve bringing* in an external consultant.
- Shall we *consider expanding* into China?
- I *suggest planning* two years ahead.
- You cannot *avoid taking* risks altogether.
- This *risks spinning* out of control.

- I'm *looking forward to seeing* you.
- I *apologise for being* late.
- We *are thinking about investing* in oil shares.

Useful vocabulary

to run through it: noch einmal durchgehen
premature: voreilig
in a nutshell: kurz und bündig
to gauge [geidʒ] the success: Erfolg beurteilen, bewerten

Diplomacy and politeness

It's bad news, I'm afraid ...

Diplomacy and politeness are not only important elements when it comes to expressing agreement and disagreement in English-speaking meetings. They also come to the fore when English-speakers are about to deliver bad news or present information that they know may not be well received by the listener.

Example: using language to persuade

A: Right, we've got some broad agreement on the way forward, but let's now focus on some practical details. Karl, do you think I could ask your team to do the figures for the presentation?

B: Well, in principle, yes, but this comes as a bit of a surprise.

A: I see. What do you think would be a reasonable deadline?

B: I'm afraid I can't promise anything right now. Shall we say by the day after tomorrow, around noon?

A: That could be cutting it a bit fine – ideally we need the figures tomorrow at 5.00 p.m. at the latest.

B: Let me think. I was wondering if we could speed up the process by outsourcing some of the graphics. Wouldn't it be faster if the charts were done by our colleagues in Prague? You know, we're all pretty busy with the Wang project, which is already slightly behind schedule.

A: Okay, fair enough. Could you get on to them right away?

Useful phrases

- I'm afraid it's not good news.
- As (I'm sure) you appreciate, this is a difficult situation.
- I'm sorry to say that ...
- Regrettably, we ...
- Unfortunately ...

Polite questions

However, there is more to diplomatic language than just lists of readymade phrases: expressing the matter in a different way, i.e. using little twists with language can help. If you ask a question, eg, instead of making a statement, your request will sound more sophisticated and less dogmatic; "would" and "could" also sound far more polite than "can" and "will".

For example, compare, "Wouldn't it be a good idea to deliver in two weeks?" with bluntly stating: "We will deliver in two weeks." And the longer a sentence is, the more politely it is perceived to be by the listener; eg "How old is the car?" sounds much better if it is preceded by an introductory phrase:

- I was wondering how old the car is?
- Do you happen to know how old the car is?

- Do you think you could tell me how old the car is?
- Would you be able to tell me how old the car is?

Checklist: sounding more polite

This table summarises what you can say to sound more polite:

What you want to say/do	What you say
Win time, make your counterpart curious about what you are going to say or prepare them for disagreement	Well actually right in fact to be honest
Give a negative statement	I'm afraid
Use a question instead of a statement to make it sound like a mere suggestion; a negative question sounds even more negotiable	Is Monday next week okay for you? Isn't Monday next week good for you?
Sound softer when refusing: "would" instead of "will"	That would be a problem.

What you want to say/do	What you say
Soften criticism – use words (qualifiers) such as "very", "slight", "some", "little", "a bit", "some"	This sounds like a slight problem. There are some reservations about the concept.
Make counter-suggestions using comparisons or negative questions	It may be more convenient to ... Wouldn't Friday be better?
Use your voice for stress	Of course we *can* do this. This is quite a *large* sum.
Sound less pushy by using the continuous form	As I was telling you ...
Avoid negative adjectives, even if they are linguistically correct	The idea doesn't seem to be very helpful. I'm not very happy with that.

A diplomatic game of give and take

Discussion partners in English-speaking cultures are often more open to acknowledging their own mistakes if the other party is also willing to admit their share of responsibility for an error. The pattern that often emerges is an acknowledgement of responsibility followed by a polite request for one's discussion partner to do the same.

Example: a little diplomacy goes a long way

A: As Rob said, we're unhappy with the quality of the goods you supplied and will be looking for you to make some kind of price reduction. I'm sure you can appreciate that we haven't been able to charge our customers full price for the goods.

B: I'm afraid that will be quite difficult for us at this late stage. We're prepared to take on board that some quality issues arose at the production stage. However, having carried out an internal investigation, we do feel that some of the features you point to were not adequately defined in the specifications you supplied to us. Would you be willing to concede that these points were not made clear to us from start?

A: With the benefit of hindsight, we can see that the specifications were not as clearly formulated as they should have been. But this doesn't change the fact that our tools division incurred substantial losses last year due to the substandard quality of the products you supplied.

B: Your business is very important to us, but unfortunately we just aren't in a position to offer a price reduction now that you have taken ownership of the goods. Could you perhaps see your way to considering a discount on your next order with us in lieu of a reduction?

A: That sounds like a workable solution. I'm sure it would go a long way towards soothing our Finance Director's headache.

B: Good. I'm very pleased to hear that.

Useful phrases

Acknowledging faults

- We readily accept that some of the errors were due to a fault in our system …
- I'm very sorry to say that we made a mistake with the order.
- As we've established, we need to address some serious issues in our production.

Asking for an acknowledgement

- Could you perhaps see your way to accepting that there's also some room for improvement at your end?
- Would you be willing to accept/concede that there were also some issues at your end with regard to ...?

Useful vocabulary

to come to the fore: ins Blickfeld geraten
cutting it fine: sich wenig zeitlichen Spielraum lassen
curious: neugierig
stress: Betonung
reservation: Vorbehalt
to incur losses: Verluste erleiden
with hindsight: im Nachhinein
in lieu of: anstelle von
to soothe: beruhigen
to stand sb's ground: sich behaupten

What to do in case of language problems

There is far greater potential for misunderstandings when you are attending a meeting that is taking place in a foreign language. Perhaps you just didn't hear what a person said or their accent is difficult to understand. Or maybe differences in the ways people from other cultures express themselves can lead to confusion. It may also happen that your counterpart uses a word or expression you have never heard in your life.

Example: asking for repetition

> A: We are talking about one billion consumers in China.
> B: Sorry, how many did you say? One million?
> A: Oh, no. One billion, of course.

In any case, you should not hesitate to ask your counterpart for repetition and clarification – there's no shame in doing so. Just use the methods and phrases given on pp. 79–83 under "Enquiring and resolving misunderstandings", namely

- asking for repetition,
- summarising for clarification,
- recapping and confirming.

Voting

Sometimes attendees cannot agree on an outcome. It may then be necessary to take a vote, which is usually done at formal meetings. However, voting should be seen as a last resort, as it will leave a number of attendees dissatisfied with the outcome. If a vote is necessary, the chair should keep strictly to the formal procedures.

A vote can either be done by secret ballot or by a show of hands. Before that, the subject of the voting has to be made clear. A suggestion or an idea that is to be put to a vote is called a "motion". Before a vote can be taken, a motion needs to be "seconded", i.e. supported, by another person. When a motion is put to the vote and agreed on, you say that it is "carried". When there is no agreement, it is "failed". Usually, majority votes are taken. In case of a tie vote, the chairperson often has the deciding vote.

Of course, the outcome of the vote has to be recorded in the minutes, eg: "Motion to allow for flexitime, moved by Peter" or "Motion to allow for flexitime, seconded by Jane."

Useful phrases

- Can I ask for a show of hands, please?
- All in favour?/All opposed?
- Those for/against the motion, please?
- Aye! [say "aye" or raise your hand to show you agree]
- Any abstentions?
- The motion was carried unanimously.
- The motion has been rejected by four votes to two.

Useful vocabulary

flexitime: flexible Arbeitszeit
to make/second a motion: Antrag stellen, unterstützen
to table/introduce/present a motion: Antrag einbringen
to carry a motion: Antrag annehmen
a motion fails: Antrag fällt durch
majority vote: Mehrheitsabstimmung
tie: Stimmengleichheit
aye!: ja

After the meeting

After the meeting is before the next meeting – to put it simply. Most meetings are followed up with a written record of what was discussed and agreed: the minutes. These then need to be passed on to the attendees.

In this chapter you will find out how to

- make the minutes (page 94),
- follow up on the meeting and take the next steps (page 99).

Making the minutes

The minutes have to be accurate and clear, summarising what was said. Lengthy sections can be boiled down to their essence. The minutes are usually circulated to all participants within a few days of the meeting and after approval by the chair. Remember that it is not necessary to include every word that is spoken, only important points and any votes and results. Indicating who said what is also necessary, which is why the minute-taker should make sure they know the names of the attendees. It is also recommendable to type out the minutes immediately after the meeting – this way you can make sure that you don't forget what was said.

Examples

Example 1: short form for minutes
Minutes of the Steering Committee Meeting

Date:	1 February 2009, 7.00–8.40 p.m.
Venue:	San Siro Meeting Room
Present:	John Snyder, Bill Meyers, Lisa Förster, Annette Joyce, Britta Pocklington
Apologised:	Anita Ferrarotti

1. New company brochure
There was positive feedback on the first revamped edition. There were 5,000 copies printed, more are needed next time. More photographs would enhance the overall appearance. Bill to contact PR for further action.

2. Decision-making process
In Greg's opinion, decisions taken by the steering committee were not valid unless the Project Leader and Management both agreed with the decisions. The other project members see this differently.

John pointed out that in all other departments the Project Leader was simply a project member with an allotted task.

Lisa suggested the matter be clarified by a member of the board. The committee decided by four votes to one to have Dr Meyers clarify the issue. Britta asked Greg what he was going to do about his position. He said he would think it over and submit his reply in writing.

3. AOB

Interest was shown in a new approach in marketing as briefly explained by John. To be followed up in the course of the next meetings. Thanks expressed to Britta for doing an excellent job during the last trade fair in Berlin.

Next meeting: To be announced.

Example 2: action minutes

Minutes of the Steering Committee Meeting

Date: 1 February 2009, 7.00–8.40 p.m.

Venue: San Siro Meeting Room

Present: John Snyder, Bill Meyers, Lisa Förster, Annette Joyce, Britta Pocklington

Apologised: Anita Ferrarotti

1. Newsletter

BM reported that our newsletter was in high demand. However, he felt that it lacked conciseness and that a more modern look would be appreciated by the readers. The discussion ended with the general decision to look into the costs for a revamp.
Action: BM by 19 Feb.

2. Digital cameras

JS suggested our staff's mobile phones should be equipped with cameras to allow for faster transmission of on-site findings. The committee members were not sure if this was necessary.
Action: All – decide by 20 Feb.

3. Floor plan – hot seating

LF proposed changing the seating arrangements, as not all staff are present in the office Mon–Fri.

AJ added that the new home office day would also make for more desk space. Suggestion welcomed by committee members.
Action: LF design new floor plan by 19 Feb.

4. AOB
JS reported on his trip to Brazil. He made substantial progress with the authorities and will present the outcome at the next meeting.
Action: JS presentation 20 Feb.

BP reminded all to come up with ideas regarding give-aways for the trade fair in Zurich in May.
Action: All by 20 Feb.

Meeting adjourned at 8.40 p.m.
Next meeting: February 20, 2009
Venue: San Siro

Tips for minute-taking

Getting prepared

If you work on a laptop computer, prepare a file containing the names of the participants and the items on the agenda. Then either take the laptop with you to fill in the key words during the discussion (to be formulated in full later on), or write them down on a sheet of paper and complete the minutes after the meeting.

Very often, initials are used to refer to participants who made a contribution or who are to carry out an action point. If you generally find it hard to remember people's names, make a habit of taking a brief note of their seating position during the introductions at the beginning of the meeting. It is best to write the names on a piece of scrap paper in the seating order. Use the attendance list to check if all the names are spelled correctly.

Writing style

As a rule, the style is impersonal and concise. There are two ways of writing the minutes: one follows the chronological sequence of what was discussed, the other sticks to the written agenda (see examples, pp. 92–94).

It is a good idea to stick to the format, style and content of the minutes which were written for previous meetings. Every organisation has its own conventions.

Variations for the word "say"
A repetition of "he said" sounds boring after reading it for the third time, so try some variation using the following verbs: mentioned, explained, confirmed, agreed, suggested, proposed, asked, introduced, discussed, reported, reminded, read out, indicated, pointed out.

Checklist: the minutes

In general, the minutes contain
▪ place and date of the meeting,
▪ names of participants: present (also when they left, if they leave early) and absent,
▪ subject of the meeting,
▪ approval of the last minutes,
▪ items on the agenda: discussions, outcomes, action items, who they are assigned to, deadlines,
▪ any other business (AOB),
▪ date, time and place of the next meeting.

Useful grammar

The minutes are generally written in the past tense. If reference is made to the future ("will" or "going to") or the present, these tenses may also be used.

Another convention is to use the passive:

▪ No extra expense to be incurred without prior consent by the board.

▪ Session to be coordinated by John.

However, it is modern style to substitute as many unnecessary passives as possible with an active sentence structure. "It was mentioned by John that ..." sounds better if it is changed into an active sentence: "John mentioned that ..."

Useful vocabulary

approach: Ansatz
tangible: greifbar
shorthand: Stenografie
allotted: zugeteilt, zugewiesen
to stick to sth: sich halten an
revamp: neue Aufmachung
costs are incurred: Kosten entstehen

Following up the meeting

After the meeting, each role has its own duty to fulfil.

- The minute-taker has to get back to the attendees with the minutes in order to communicate the allotted tasks to the persons concerned.
- The chair has to follow up to ensure that all the agreed action items are carried out.
- The owners of action items should
    - complete them asap,
    - report back as agreed,
    - liaise with others, if necessary.

Example: email with attachment

Subject: Minutes of the Steering Committee Meeting
Dear all,

I am writing to thank you all again for the fruitful meeting last Tuesday. Please find attached the minutes as well as the updated contact list.

Could you please get back to me asap regarding the time and date for the next meeting? John suggested the 20th at around 7.30 p.m. Please let me know if this is convenient.

I look forward to seeing you again soon.

Best regards,

Lisa

Useful vocabulary

scrap paper: Schmierpapier
asap (= as soon as possible): so bald wie möglich
to liaise: Kontakt aufnehmen

Special types of meetings

With the advent of new technologies, new types of meetings, such as videoconferencing, became possible and more widespread. Project management, too, has given rise to new categories of meetings, as have creative techniques. Negotiations and customer meetings have become more global in terms of attendance, requiring greater cultural awareness and thought than meetings with your fellow nationals.

This chapter helps you find out about the particularities of

- meetings with customers (page 102),
- negotiations (page 105),
- briefing and brainstorming (page 111),
- jours fixes and kick-offs (page 114),
- telephone conferences (page 116).

Meetings with customers

Getting in touch

Whether you meet the customer at their offices or at yours, make sure you make an appointment well in advance. Be prepared to offer several alternatives to suggest a meeting time that is really convenient for your (prospective) customer. Moreover, try to use the channel of communication that seems to be most appreciated by the other party: some people hardly ever check their emails, while others hate being disturbed by phone calls. But, first of all, you have to get in touch with them. When doing so, try to point out the specific issue that connects your company with them.

Useful phrases

- We used to have a branch office in Hamburg, close to your headquarters in the city centre.
- I learned from your website that you also work with H&Z. We've been their preferred suppliers for eight years now.
- We did a similar project for ABC, Inc. three years ago.

Identifying your client's needs

Once you have managed to get an appointment for a meeting, what counts is responding to a client's needs. First you need to know what these needs are by finding out how your products or services meet the customer's requirements. If they don't meet them yet, think about how you could bring this about. Find examples among your references to show that you have already met similar needs in the past.

Useful phrases

- So you said you needed a new training partner for internet applications?
- When will you open the new plant in Hong Kong?

Explaining your proposal in detail

Another crucial point is to create rapport with your customer, especially if this is a newly acquired customer. Apart from making small talk, you should establish common ground by telling them why your two companies fit together.

Useful phrases

- Did you see that our connector uses wireless technology?
- Can I draw your attention to the fact that fuel consumption is only three litres for this model?
- If you look at the illustration, you can see that this is where our product can be switched to 110 Volt. We use our own patent.

Anticipating objections

Any potential new customer will check out very carefully if you are the right person or company to meet its needs. Anticipate any concerns they may raise and address them by using positive statements.

Useful phrases

- We do have certificates according to DIN ISO for the product.

- An export certificate is not required for shipments to Bali.
- You don't need to worry about different sockets, we build in the British version as standard.

Ending the visit

It is highly recommended that you save some time at the end of the meeting to address any questions. Even if you cannot give an answer off the cuff, you can make a list of action items to follow up by email or letter. And of course, don't forget to use friendly parting phrases to end on a positive note.

Useful phrases

- Are there any questions I could help you with?
- Could we perhaps discuss delivery periods next time? And then there is still the issue of transportation costs.
- Would next week suit you for another meeting?
- I'll email you the documents as soon as I can.
- It was a pleasure meeting you.
- Thank you for coming. It's been a fruitful meeting.
- Have a safe trip home!
- I'm looking forward to meeting you again next month.

Attentive hosts

Being an attentive host is most important in a customer meeting. Apart from small talk (see p. 40), this involves watchfulness and attention, eg by providing ice cubes for your American guests. Preparing an agenda for the meeting and making it available to the client beforehand will be appreciated and show that you really care about this sales contact, just as much as a thank-you letter after the visit, accompanied by the meeting minutes.

Useful vocabulary

crucial: wichtig
fuel consumption: Kraftstoffverbrauch
off the cuff: auswendig, ohne nachzuschlagen

Negotiations

Everybody negotiates. On a day-to-day level, just as when the stakes are high in an international merger, wage negotiations or for a customer-supplier contract. Apart from the principle of win-win, all sorts of tactics are employed deliberately. The most important thing, however, is to know what you want to walk away with. In general, the following tips apply to all types of negotiation:

Checklist: negotiations

- Ask a lot of questions and listen, listen, listen. The more you find out, the better ("Active listening", p. 67).
- Use diplomatic language whenever possible.
- Make sure you are understood correctly.
- Make it clear that you understand what the stakes are.

Example: price negotiation

A: So, Will, how much do you have in mind, let's say per unit?

B: Well, it all depends on how many units you would like to sign up for on a regular basis.

A: Sure, but I would really need to hear a basic price from you first: I need to see if we are thinking along the same lines.

B: Okay, as you may remember from our written proposal we were thinking about 350 dollars per shipping unit.

A: 350 dollars! Are you kidding? I could get each pineapple wrapped in gold foil from the Ivory Coast at that price! And they would arrive in Europe much earlier, too!

B: But you know how the market is at the moment. Supply is at a premium. We have reserved 300 units for you because you are one of our long-standing customers. Otherwise we could have sold them already, and for a much higher price than that.

A: Okay, well, I will have to contact head office about that.

B: Certainly, there's no hurry, but don't forget our quality and reliability when it comes to delivery. Our fruit has always been good value for money.

A: I know, Will, I know. But what if the fruit goes bad during transit or if a container gets damaged? We would require insurance covering such events.

B: Of course we would grant you compensation, no question about that. The sum of the actual damage would have to be determined by an independent expert, however, at your expense.

A: If you don't mind, we'd better discuss this point next time. We can't move on this right now. And I feel quite tired after the long flight. I guess we should call it a day.

B: You're right. I want to see you in good shape tomorrow for the tour of the plantation.

A: So, before we meet next time, I'll find out about order quantities and you look into the point about shipping.

B: Right, and the insurance matter and the other outstanding issues as well.

A: Fine, so that's it then. Where are we going for dinner, Will?

> **Right contact and enough time**
> On an international level, it is important to invite or to send the right person to a negotiation. In Japan, a young manager might not be accepted, even if he or she has decision-making powers. Also bear in mind that other cultures need more time for decision-making and attach more importance to establishing a good relationship first. Some will also only do business with a person they trust.

Useful phrases

Establishing common ground and reformulating

- You know, the plot of land is worth a million right now.
- The track record of your module is really impressive.
- Let me check if I understood you correctly. You said you were looking for a long-term supplier?

Exploring positions

- Could I ask you how much you had in mind?
- When exactly do you need the material?
- Can you tell us what your standard terms are?

Making suggestions

- If you place an order for more than 5,000 units by the end of the month, we will ship them free of charge.
- We may agree to your terms, provided that you'll give us more leeway with regard to delivery times.
- It depends on how much you would be prepared to pay for our service.
- I'm afraid we can't accept that unless you offer us a three-year warranty.
- How about 50 pieces?

- Well, actually, 55 would be better.
- Why don't we try a more middle-of-the-road approach?
- Alternatively, we could offer you a discount.
- Let's think about preferred supplier status first.
- What if we offered you five per cent?

That might/may/would/could/can be an option!

Modal verbs are indispensable in negotiations. You can use them to vary the degree of certainty of your statements, i.e. you make what you say more or less probable.

- We *can* guarantee you delivery in five days. [Fact]
- We *could* deliver next week. [Real possibility]
- We *would* guarantee delivery ex works. [Possible, but under certain conditions]
- We *may* guarantee delivery to Bremen. [Maybe]
- We *might* deliver to the Arab Emirates. [Faint possibility]

Softening disagreement

- I'm afraid we really can't agree to five per cent.
- Unfortunately, that's not really the way we see it.

Pushing for a decision

Soft-sell approach

- We would need to see some movement on price.

Hard-sell approach

- Take it or leave it!
- This is our final offer!
- We'll have to call the whole deal off/take our business elsewhere.

Refusing

Soft-sell approach

- I'm afraid that's not quite what we had in mind.
- I'm afraid this is as far as we can go.
- I am very sorry, but your offer still does not convince us.
- We feel that this is a bit much.
- I'm afraid I'm not in a position to grant you that.

Hard-sell approach

- No way! You know what the market is like right now.
- That's completely out of the question.

Asking for more time and looking ahead

- I'll have to think it over.
- I'll have to talk to my line manager about that.
- We can't give you a definite answer just now.
- As a next step we should look at ...
- If you don't mind, I'd like to come back to that later.
- Let's not rush things.
- I don't think we should make a decision just yet.

Agreeing

- That sounds/seems reasonable.
- That sounds like a sensible suggestion.
- In fact, that suits me fine.
- It's a deal!

Summarising what was agreed

- Can we run through what we've agreed?
- So, I'll summarise the important points of your offer.
- I'd like to check/confirm what we've said.

Next steps

- We need to meet again soon.
- So, the next step is to draft a formal contract.
- Before the next meeting we'll check the order quantities.

> **Follow-up**
> A written follow-up summarises what was agreed and obliges the other party to act. It also shows that your intentions were serious. Sometimes a formal "letter of intent" is sent before any contracts are signed.

Useful grammar

How to express a must

Don't forget that "must" has no negative form and exists only in the present tense. "Must not" means "not allowed to" and, as it can only be used in the present tense, for all other tenses you have to resort to "have to do sth" in its respective tenses:

- We *must* meet again to discuss the details.
- We *mustn't* forget to send the draft today.
- We *had to* organise the TC differently.

If-clauses

An essential tool for bargaining are if-clauses. Particularly type I and II prove to be useful.

- Type I if-clause: The if-part of the sentence is in the present tense, the second part is formed by "will" + infinitive. It is used to talk about real possibilities: "If you agree on that, we will offer you a discount."

- Type II if-clause: Type II is used to express a hypothetical possibility. The if-part of the sentence has to be in the past tense, with "would" + infinitive in the second part: "If you agreed on that, we would offer you a discount."

All if-clauses can also be turned around, so that the if-clause forms the second part of the sentence, eg: "We *will offer* you a discount, if you *agree* on that."

Useful vocabulary

what the stakes are: was auf dem Spiel steht
plot of land: Grundstück
leeway: Spielraum
on-site maintenance: Wartung vor Ort
line manager: unmittelbarer Vorgesetzter
at your expense: auf Ihre Kosten

Briefing and brainstorming

What these two types of meetings have in common is that they require a subtle and careful facilitation style. In brainstorming sessions the chair should give the participants the opportunity to have their say and interrupt as little as possible; in briefings he or she should make sure all the information comes across.

Briefings

As the name suggests, a briefing should be kept short by definition. It should not take longer than 30 minutes. In a briefing the information has centre of attention. The idea is to share and spread knowledge. It is a one-way transfer of information from the organiser to the attendees. There is no exchange of ideas, nor are decisions taken. Questions may be asked by the persons being briefed.

Example: briefing meeting for a facilitator

A: Thanks for coming in, Jane. Have you received the guidelines for the first group?

B: Yes, I have. In fact, I've read through them and I'd like to ask you some questions. As always, the devil is in the detail!

A: That's excellent, but first let me show you how the device works. Then I'd like to tell you what our main focus is.

B: Good, please go ahead. How do you switch it on?

A: The buttons are on the back panel, all of them.

B: I see. And what shall I say if people ask me who the sponsor of the study is?

A: Just tell them you don't know yourself, and that it is one of the major players in the market.

Useful phrases

- This is what we want to achieve: increased timeliness and accuracy.
- Just show your customers this list.
- There's no need to feel uneasy.
- Are these figures understandable?
- Do you understand what I am getting at?

Brainstorming sessions

Brainstorming is a creative technique aiming at the maximum number of ideas to address a certain area of interest or to solve a problem. In the first step, the value or feasibility of the ideas generated is not an issue – all ideas are welcome. In general, brainstorming makes most sense for group sizes of 4 to about 20. Superiors or managers should only be invited if their subordinates will not feel inhibited by their presence. There should either be a time set for collecting ideas or a maximum number of ideas (50 to 100) fixed in advance. After that time, the ideas are categorised and sorted. But first of all, the topic of the brainstorming session has to be set out clearly. Furthermore, it is important that all participants make an active contribution. As a rule, criticising ideas is not allowed, as this might hinder inventiveness and spontaneity. It may sometimes be necessary for the chair or the facilitator to remind participants of these basic rules.

After the time set for collecting ideas is over or when there are no more ideas coming from the participants, the suggestions are sorted and categorised and any unfeasible ideas are deleted from the list. This can be done either as a group or by asking each person to pick out the five ideas he or she likes best.

Useful phrases

- So, how can we achieve better customer service?
- What aspects can enhance the overall customer experience?
- Can we agree not to interrupt each other?
- Can I ask you to save your comments for later, please?
- Let me remind you of the basic rules of brainstorming …

- Can you please let Peter finish what he was saying?
- The following phrases may come in handy when taking notes on a flipchart during brainstorming.
- What was that again?
- Sorry, could you repeat/spell that for me, please?
- I think we have a double entry here.
- I see two main directions here, would you all agree?
- Can we filter out all the pros and cons, please?
- Shall we organise these ideas into two or three columns for a better overview?
- Can each of you now select the five ideas you think are the best?

Useful vocabulary

sponsor: Auftraggeber
feasibility, feasible: Machbarkeit, machbar
inhibited: gehemmt

Jours fixes and kick-offs

Jours fixes

The term "jour fixe" is derived from project management. The same day of the week and time is reserved for an internal meeting with the same people, eg the first Wednesday of a month or a week. The aim is to facilitate the exchange of information and to update all project members so they are at the same knowledge level.

Example: jour fixe team meeting

A: ... to put you all in the picture and bring you up to date with the latest developments.

B: Thank you, Brian. Are there any other points to be discussed?

C: Well, yes, actually, but it's a rather delicate issue.

B: Go ahead, Valerie, we're all curious.

C: Erm, right. It's about our colleague, George. He always passes the buck. He calls in sick every time a deadline is due.

A: I must say that I also find him very rude. [All the team nods]

B: I see, I see, so we seem to have an issue with George ...

C: An issue! Are you kidding? I'm fed up with being on the same team, let alone in the same room with this guy.

B: Val, just try to take it a little easier on him.

C: Why are you defending him like that?

B: Well, I'm not supposed to tell you this, but George isn't too well at the moment. He was diagnosed with a serious illness and is currently undergoing treatment. This is why he is often away from work.

C: Oh, I'm so sorry, I had no idea!

A: That's a different story. Poor George!

Regular team meetings increase commitment and prevent misunderstandings and rumours. Apart from conveying information, it is important to foster the feeling of being appreciated among the team. This is why the participation of all team members should be encouraged. They should always be addressed by name, including when giving positive feedback: "Thank you, Damian, for bringing this point up."

How to prevent boredom
Variety can be added by altering the style in which the meeting is held, as well as the activities that take place. Try to encourage all participants to become actively involved rather than just listening passively.

Kick-off meetings

Kick-off meetings are extremely important for the success of a new project. Their purpose is to signal to all team members and stakeholders that the project has now begun. They give the team orientation and direction and ensure that everyone understands what their roles are. As far as language is concerned, kick-off meetings contain the following features:

- introducing people (which is an important feature of a kick-off meeting, in particular when project owners and stakeholders have never worked together before),
- giving information (on the purpose of the project, its major deliverables, its risks and assumptions, as well as its milestones),
- getting agreement (on suggested project management procedures from all stakeholders),
- asking and answering questions.

Useful vocabulary

to facilitate: moderieren
to pull one's weight: seinen Beitrag leisten, sich ins Zeug legen
to pass the buck: den schwarzen Peter zuschieben
rude: unhöflich, barsch
stakeholder: Prozessbeteiligter (eigentlich: Aktionär)

Telephone conferences

Telephone conferences (TCs) are increasingly widely used. Important factors for the success of a TC are a clear agenda, as well as the discipline of the participants and unambiguous communication. The agenda should contain the timing for individual points and the allocation of roles, such as the chair, timekeeper, secretary and the call facilitator, who takes care of the technical aspects.

Agenda

Example: subject of a telephone conference

Agenda: outsourcing event organisation

Date: 5 May 2009, time: 11.00–12.00 CET

Participants: Dörte Gluchowski (chair), Peter Pan (call facilitator), Annette Joyce (timekeeper), Lisa Förster (secretary), Bridget Mayer, Marén Volkers, Tanya Feldman

Item	Presented by	Time
1. Apologies	DG	5
2. Minutes of the previous call	LF	5
3. Approval of agenda	DG	5
4. Report on Hamburg trade fair	MV	10
5. Outsourcing proposal – discussion	AJ/all	25
6. Date for next conference call	DG	5
7. Feedback – discussion	DG	5

For a larger number of participants, it pays to prepare the agenda carefully and to circulate it in good time. The agenda should contain the timing for each point. When calculating the timing, you should bear in mind that each person's active contribution will take about three to five minutes per item. In our sample agenda, we have seven participants and 25 minutes allocated to the main discussion item. On the whole, the agenda should be kept as short as possible in order to limit the length of the meeting, as a TC is far more demanding in terms of concentration than a face-to-face meeting. If the TC is only held between a limited number of people and on a routine basis, an agenda is sometimes just made ad hoc, that is, at the beginning of the call.

Useful phrases

- Can we just quickly make a list of the items we need to talk about today?

- Does anyone have an agenda or do we need to jot a few items down so that we don't forget anything important?

Starting a conference call

Before the serious part of the conference call, participants are encouraged to make small talk, just as if they were meeting face to face. On the one hand, this will create better rapport between them, just as if they were in a real meeting. On the other hand, they can all tune in to each other's voices and accents – especially important for non-native speakers of English. Then, the chair starts the TC by welcoming the participants, doing a roll call, reviewing the agenda and summarising the aim of the TC.

Useful phrases

- Hi, this is Dörte. I'll be the chair in today's conference call.

- Can we go round the table and hear who has logged on already? Can you just say "yes" when I call out your names?

- Before we start the ball rolling there are a few technical issues to clarify. Have you all received the agenda?

- Are there any other items we need to talk about today? Or any suggestions regarding the agenda?

- As you know, the main focus will be on the outsourcing business. It's important that we establish common ground on further procedure.

The chair should also clarify if participants need to leave the call early:

- Does anybody have to leave the call early?
- Yes, I do, actually. That's Pauline.
- Pauline, when do you have to log out?
- Around twelve at the latest.

By the same token, participants should announce if they are taking a break and let the others know when they are back in the TC:

- Sorry. It's Lisa. I need to print out the agenda again. I'll be back in a minute.
- Annette here. I'm back in the call.

A memorable self-introduction
If the participants do not know each other, they should introduce themselves so that the others know who the new person is and why they are joining the TC. They should say their names and what they would like to be called, what their job involves plus a past achievement and something personal, as well as a "memory hook" so that the others will remember them better.

Controlling the meeting

The chair has to involve all participants and ensure that they take turns. On the other hand, it is sometimes necessary to cut a speaker short. In any case, the role of timekeeper is vital for sticking to the agreed timings. He or she should remind everybody to keep it short. In longer TCs it sometimes makes sense to interrupt the call for a few minutes to give everybody a chance to collect their thoughts, to find some more information or simply to stretch their legs.

Creating rapport

If you want to stress the group feeling and convey an atmosphere of co-operation, you should use the we-form. It includes everybody in the group and helps create rapport with the participants. That's why you should say "*We* think this is a viable suggestion" instead of "*I* think this is a viable suggestion".

Example: directing a TC

A: Bridget, would you kick us off, please?

B: Well, this is a great way to cut a lot of unnecessary spending. My figures show that we can save a lot of money.

C: Can I just come in here? That reminds me – we can't ask Marketing to organise the event again. It's unfair.

A: Sorry, Tanya, but I think this is a bit of a sidetrack. Let's try and keep to the agenda, okay?

C: Of course.

B: As I was saying, outsourcing the event organisation can really save a lot of money. It's obvious, if you ask me.

A: What do the others think? Is there anything else we should consider? Nothing? Okay. We seem to have dealt with the outsourcing issue. Let's move on to the website. We've budgeted 6,500 dollars to redesign it. Peter, what do you think?

D: Well, obviously we need to make cuts and this would be less painful than many of the others. But we really have to prioritise PR because this is the bread and butter of our business. Perhaps we can consider a limited redesign?

C: I agree.

B: I do, too. You know, Peter, speaking of the website, you should really consider using a different designer this time. I don't think the last one did a good job at all.

Useful phrases

- Bridget, what do you think?
- Perhaps Maria could make a start with her opinion on item six?
- I need to hear your opinions about the new structure. Can we go round the group quickly? Franz?
- Thank you, Elke – Mark?
- Marty, I think Bridget wanted to add something? Go ahead, Bridget!
- Sorry, Toni, but I don't think this is really our topic today. Can I ask you to come back to the issue at hand?
- Sorry, Lisa, but we said five minutes only for the review of the minutes.
- You've only got three minutes left, Marén. Could you speed your report up a little, please?
- Dörte, sorry to interrupt, just to remind you that we only have ten minutes left to wrap up the call.
- How about taking a ten-minute break here? Shall we reconvene at 12.45?

Documents are usually distributed by email before the TC. If you need to refer to them, you can use the following phrases:

- Could you all turn to page three, please?
- Have you all got a copy of the contract to hand?
- You'll find the total sum at the bottom of the page, on the right/in the bottom right-hand corner.

Ending a telephone conference

Wrapping up a TC involves the same elements as a normal meeting: summarising results, allocating action items and defining the next steps, including a time for the next meeting (see p. 57).

Useful phrases

- Let me sum up the arguments for and against outsourcing.
- If I understood right, most people were in favour of Bangalore as the location for our new subsidiary.

Feedback on the TC

If telephone conferences are held on a regular basis, it is useful to ask the participants for their feedback on how the TC worked for them. The results can then be discussed at the beginning of the subsequent meeting or beforehand by email. Improving how TCs are run will ensure their ongoing success. The feedback should cover the following questions:

- Did the technical equipment work well?
- Were the timings adhered to?
- What could be done better?
- What should we continue to do and what should we stop doing?
- What should be avoided?

Checklist: TC dos and don'ts

- Introduce yourself the first time you speak: "Hello, everyone. My name is ... and I work in the ... department."

- Make small talk to give the other participants a chance to "tune in" to your voice and way of speaking: "How is the weather over in California? It's rainy here in Munich."

- When introducing yourself, mention something personal plus a memory hook so that the others will remember you better.

- Speak slowly, clearly and concisely. Use your voice actively, as people cannot see your facial expression.

- Use your name every time: "This is John speaking ..."

- Describe your body language, for example: "I'm shaking/nodding my head here."

- Announce that you are leaving or returning after a break, for example: "This is John speaking, I am about to leave the call for five minutes./This is John, I'm now back in the call."

- Keep in mind that all paper rustling, sipping or clicking pens will be heard by everybody.

- If you ask somebody for their input, address them with their name, repeating the question if necessary.

- Learn how to work the equipment properly, especially the mute button: remember others might hear the hold music while it is pressed.

- Don't sit on a leather chair! The sounds it makes when you move around will be quite startling to yourself and others. A fabric-covered chair is much safer.

Literature

"How to manage meetings" von Alan Barker, Kogan Page, London, 2002

"The language of meetings" von Malcolm Goodale, LTP, Hove, 1987

"Fifty ways to improve your Telephoning and Teleconferencing Skills" von Ken Taylor, Summertown Publishing, Oxford, 2008

Index

Bibliografische Information der Deutschen Bibliothek
Die Deutsche Bibliothek verzeichnet diese Publikation in der Deutschen Nationalbibliografie; detaillierte bibliografische Daten sind im Internet über http://dnb.ddb.de abrufbar.

ISBN 978-3-448-09297-4
Bestell-Nr. 01307-0001

© 2009, Rudolf Haufe Verlag GmbH & Co. KG, Niederlassung Planegg b. München
Postanschrift: Postfach, 82142 Planegg
Hausanschrift: Fraunhoferstraße 5, 82152 Planegg
Fon (0 89) 8 95 17-0, Fax (0 89) 8 95 17-2 50
E-Mail: online@haufe.de
Internet: www.haufe.de
Redaktion: Jürgen Fischer

Konzeption und Realisation: Sylvia Rein, 81371 München
Lektorat: Barbara Imgrund, 69121 Heidelberg, Sylvia Rein, 81371 München
Umschlaggestaltung: Kienle gestaltet, 70178 Stuttgart
Umschlagentwurf: Agentur Buttgereit & Heidenreich, 45721 Haltern am See
Druck: freiburger graphische betriebe, 79108 Freiburg

Die Autorinnen

Lisa Förster

ist Übersetzerin, Dolmetscherin und Sprachtrainerin für Englisch und Französisch und mit eigenem Übersetzungs- und Sprachtrainingsinstitut selbstständig. Sie arbeitet für Unternehmen und Sprachinstitute und bietet u. a. auch Fachsprachenkurse an. Im Haufe Verlag hat sie zahlreiche, erfolgreiche Bücher zu Business English veröffentlicht.

Annette Joyce

ist staatlich geprüfte Übersetzerin für Englisch und als Lektorin und Layouterin in der Verlags- und Übersetzungsbranche tätig, mit Schwerpunkt auf mehrsprachigen Büchern und Zeitschriften.

Weitere Literatur

„Business English. Useful phrases für alle wichtigen Anlässe", von Lisa Förster, 180 Seiten, mit CD-ROM, € 16,90. ISBN 978-3-448-07551-9, Bestell-Nr. 00086

„Die besten Bewerbungsmuster Englisch", von Lisa Förster, 176 Seiten, mit CD-ROM, € 19,80. ISBN 978-3-448-08781-9, Bestell-Nr. 04078

TaschenGuides – Qualität entscheidet

■ **Der Betrieb in Zahlen**
- 400 € Mini-Jobs
- Balanced Scorecard
- Betriebswirtschaftliche Formeln
- Bilanzen
- Buchführung
- Businessplan
- BWL Grundwissen
- BWL kompakt – die 100 wichtigsten Fakten
- Controllinginstrumente
- Deckungsbeitragsrechnung
- Einnahmen-Überschussrechnung
- Finanz- und Liquiditätsplanung
- Die GmbH
- IFRS
- Kaufmännisches Rechnen
- Kennzahlen
- Kleines Lexikon Rechnungswesen
- Kontieren und buchen
- Kostenrechnung
- Kleine mathematische Formelsammlung
- VWL Grundwissen

■ **Mitarbeiter führen**
- Besprechungen
- Führungstechniken
- Die häufigsten Managementfehler
- Management
- Managementbegriffe
- Mitarbeitergespräche
- Moderation
- Motivation
- Projektmanagement
- Spiele für Workshops und Seminare
- Teams führen
- Zielvereinbarungen und Jahresgespräche

■ **Karriere**
- Assessment Center
- Existenzgründung
- Gründungszuschuss – Erfolgreich in die Selbstständigkeit
- Jobsuche und Bewerbung
- Vorstellungsgespräche

■ **Geld und Specials**
- Die neue Rechtschreibung
- Eher in Rente
- Energieausweis
- IGeL – Medizinische Zusatzleistungen
- Immobilien erwerben
- Immobilienfinanzierung
- Sichere Altersvorsorge
- Geldanlage von A–Z
- Web 2.0
- Zitate für Beruf und Karriere
- Zitate für besondere Anlässe

■ **Persönliche Fähigkeiten**
- Allgemeinwissen Schnelltest
- Ihre Ausstrahlung
- Business-Knigge – die 100 wichtigsten Benimmregeln
- Mit Druck richtig umgehen
- Emotionale Intelligenz
- Entscheidungen treffen
- Gedächtnistraining
- Gelassenheit lernen
- Glück!
- IQ – Tests
- Knigge für Beruf und Karriere
- Knigge fürs Ausland
- Kreativitätstechniken
- Manipulationstechniken
- Mathematische Rätsel und Knobelaufgaben
- Mind Mapping
- NLP
- Peinliche Situationen meistern
- Schneller lesen
- Selbstmanagement
- Sich durchsetzen
- Soft Skills
- Stress ade
- Verhandeln
- Zeitmanagement